FRUIT IN BRAZIL

SILVESTRE SILVA

TEXT BY
HELENA TASSARA

EMPRESA DAS ARTES

I dedicate this book to
Natália and Rosa Maria
in recognition of their help
and support at all times.

FRUIT IN BRAZIL

SILVESTRE SILVA

I hold in my mouth the savors
of gabiroba and jambo,
the color and fragrance of the forest,
picked from the tree. Distinctive.
Araticum, araçá,
ananás, bacupari,
jatobá... all together,
a green congress in the forest,
and each one separate,
each fruit, each flavor
a sentiment composed
of the fruits, all wild,
that I take to my mouth
as if it were the forest.

Carlos Drummond de Andrade
MENINO ANTIGO

INDEX

Abiu 34
Amora-do-mato 36
Apple 169
Araçá 21
Aroeira-vermelha 28
Assai 15
Avocado 18
Babassu 49
Bacaba-de-leque 57
Banana 41
Barbados cherry 23
Baru 51
Bilimbi 59
Biribá 54
Brazil-nut 71
Brejauva 62
Buriti 44
Cacao 94
Cagaita 78
Camu-camu 64
Carambola 87
Carnauba 81
Cashew 66
Cereja-de-cametá 104
Chestnut 74
Chichá 76
Coconut 83
Coffee 97
Cupuaçu 90
Curiola 102
Dendê 105
Feijoa 112
Fig 114
Fruta-de-lobo 110
Fruta-de-tatu 108
Gabiroba 124
Grape 227
Grumixama 126
Guabiju 119
Guaraná 120
Guariroba 128
Guava 117
Ingá-cipó 133
Jabuticaba 144
Jambolan 142
Jambu 139
Jaracatiá 136
Licuri 150
Lime 148
Macadamia 157
Macaúba 197
Maçaranduba 160
Malabar almond 38
Mama-cadela 189
Mango 166
Mangosteen 172
Maraja 180
Marmelada-nativa 182
Melon 186
Monguba 191
Monkey Pot 222
Morgote tangerine 200
Mulberry 37
Murici 204
Murumuru 202
Orange 153
Papaya 163
Passion fruit 175
Perinha 214
Persimmon 92
Pimenta-de-macaco 218
Pineapple 25
Pindaíba 216
Sapota-do-solimões 220
Sorva 225
Souari nut 207
Soursop 130
Strawberry 193
Surinam cherry 211
Sweetsop 30
Watermelon 184

"Fruit in Brazil" is a fertile and fascinating subject, impossible to exhaust, and which once again we are proud to bring before our readers in the form of this beautiful book.

With the publication of this work, we also hope to be able to contribute toward making many of these native fruits better known, and hopefully better appreciated, with their largely unexplored and flavorsome potential, and which in many cases have managed to remain hidden away in the dense forests or in the immense savannahs, for the delight of only a few people.

The Publishers

Silvestre Silva rarely stops searching, seeking out and finding, some of the most remarkably curious, obscure and little-known fruit growing in this country, bringing them back for our enjoyment, we who have lived so close to, and yet so far from, these national treasures.

Luminous spheres, strange leguminous shapes, shining, some, in the noon-day sun, others half-hidden, appearing timidly in the remote corners of tropical forests. Like a dragonfly, the photographer's gaze rests lightly upon them all, tracing the letters of a curious alphabet – the writings of a deeply national world.

Abiu, Baru, Brejaúva, Caqui, Caçari, Cagaita, Chichá, Dendê, Feijoa, Grumixama, Inajá, Macadâmia, Mangostão, Macaúba and Mucajá; Pindaíba, Sorva and Sapota-do-Solimões.

In his constant quest, Silvestre Silva has needed more wisdom than courage, and the tenacity to overcome the lethargy, loneliness and apparent monotony of a silent world, to which he is now eloquent messenger.

Helena Tassara has composed a narrative to Silvestre's photos. Based upon scientific data, she plunged into the studies of a master – Luis da Camara Cascudo and his História da Alimentação no Brasil, *digging deep into the works of Pio Corrêa, diplomat, writer and lover of Brazilian fruit – author of six voluminous and magnificently useful volumes on this theme. Lengthy conversations with Silvestre completed her quest. As a result, Helena Tassara, with the help of her husband Flávio, gave form to the face of fruit, counterpointing with informative grace the remarkable images in this beautiful book.*

Maureen Bisilliat
São Paulo, June, 1996

I will never forget Domingos, an inhabitant of the Amazonian waters. I met him by chance, on a dockside in Belém - Porto de Palha, while looking for someone to take me to an island in the Bay of Guajará, to photograph the marajá palm tree. I wanted this plant, to record forever a picture of its fruit. Luckily, Guajará was the area where Domingos survived, so he took me around by motor boat showing me the charms of his bay. Later in a fragile dugout canoe, he took me into the maze of backwaters in the island of Combu on the Guamá River. Wherever he went, if there was someone there, he would shout out his news, with uncontrollable native excitement, "He has come from São Paulo to photograph a marajá !"

Hearing Domingos spreading in his manner the news of my arrival on Combu, I recalled Piedade do Paraopeba, my mountain birthplace, today swallowed up by asphalt and real estate developments out of Belo Horizonte. While we proceeded aboard the dugout along sinuous backwaters toward my meeting with this so eagerly sought palm tree, I silently made a promise that will still come about: to return one day to those same mountains, to the mysteries of the Piedade Mountain, to embrace that bacupari tree where, 15 years before, the project was born that is now producing the second of the "Frutas Brasil" series. It is the Utopia of a dream come true, to offer to the movements resisting the destruction of forests and memory, a cultural record of the fruit of the mutilated Brazilian flora.

Over mountains and plains, along rivers and roadways, by plane, boat, car, horse and often enough on foot during frequent hikes for miles into forests, always carrying my camera, 15 years of searches and discoveries have gone by. During recent years, for the particular project of this book, I returned to searching a good part of the country. There were weeks in Amazonia, traveling the great rivers, make incursions from their banks. I saw the fruitful miracles of irrigation in the São Francisco and Açu Valleys. I weathered the dust of the arid backwoods to discover species that contribute to man's survival, such as the macaúba in Rio Grande do Norte. I passed by the farm lands, savannas, fields and cities of Roraima, Bahia, Ceará, Pernambuco, Alagoas, Goiás, Mato Grosso, Paraná, Santa Catarina, Minas Gerais, Espírito Santo and Rio Grande do Sul. I followed trails that led me to ranches in the interior of São Paulo and Rio de Janeiro. I made friends and allies, hid emotions, collected new stories that some day will be told. And I photographed continuously.

In the summary of this book, taken from thousands of accumulated pictures, over one hundred fruits have been presented, all with their respective scientific and cultural references. A few, due to their importance, have been shown again in this book, but with a different text. Another group is formed of the abundant irrigated fruit, which man's modern technology is placing in the great national and international consumer markets. We also have rare species from the **cerrado** *or savanna area, forgotten to popular memory, such as the pindaíba. And another that Brazil nearly allowed to die, the feijoa, today being reintroduced into Brazilian fruit culture. Two stories worth telling.*

The pindaíba fruit does not always appear in dictionaries. In Brazil's Aurélio dictionary, the name serves only to identify "lack of money". The real pindaíba is a red, exuberant custard apple, practically in extinction. I photographed a single tree on a ranch lost in the interior of the state of São Paulo. I arrived there thanks to a chain of clues involving four people, one of whom knew of someone who knew where there was a pindaíba. And so I managed to find this someone, who took me personally to this farm blessed with the existence of this rare tree.

As for the feijoa, it is a native fruit that has almost disappeared in Brazil. No one knows how, but it reached New Zealand and Mexico, who export it to various countries. It is a delicious fruit, similar to the guava, slightly more oval in shape. It has an exceptionally beautiful flower and is used as an ornamental plant. Now the feijoa is back in Brazil, thanks to an agriculturist in the town of Limeira, São Paulo, Sr. João Ernesto Dierberger, who imported seedlings from New Zealand and is cultivating them on his specialized nursery garden farm.

In spite of the genuine pleasure of contributing to this volume of "Fruit in Brazil", the project however continues unfinished. It is necessary to proceed, with many lenses, along the pathways of Brazil capturing the agonies and joys of our flora.

And I shall proceed, for two reasons. The first because, as a photographic artist, I depend upon the signs of life and knowledge that I manage to capture. When, for example, I encounter people such as Sr Expedito, of Macaubal, Rio Grande do Norte, an interlocutor with nature, he says of the macaúba, which is his food, "One eats a few of these little fruit, drinks some good water and makes it through the day".

The second reason is tellurian. Like a good son of Minas Gerais, I always want to know what's on the other side of the mountain.

Silvestre Silva

THE AMAZONIAN ORCHARD

In Amazonia, the cultivation of native fruit trees began with the aborigines, who would grow a few fruit trees from the surrounding forest close to their dwellings. In many cases these small orchards had profound religious connotations and a great conservationist meaning, as it was the belief of the Indians that a beautiful and tasty fruit was a gift of the Gods, and as such should be cared for and shared.

The colonization process brought many other fruit species from other regions in Brazil and from other continents, further enriching the already notable Amazonian orchard. Many plants found the environment in the Amazon region so favorable that it became difficult for the layman to determine their origins. Such is the case of the Mammee Apple, which in Brazil is known as the abricó-do-pará *or Pará apricot, as it is so abundant in that state; the Madagascar plum too, in Amazonia, is popularly known as the Cametá cherry in allusion to that city where it is mainly to be found; the mango generated the toponym of "Mango City" for Belém, Pará.*

Amazonian writer Leandro Tocantins brilliantly summarized the exceptional diversity of the fruits in this region in his book "O rio comanda a vida" *(The river controls life) when he says* "in this garden of Hesperides the most exacting pomologist will find the fruit that suits his palate".

A definition of the number of species bearing edible fruit in Amazonia is still a challenge for botanists and scientists. Some state roughly two hundred species with a potential for new cultures. Others think there could be more, but no one dares to quote an exact number - this is a secret held by the Gods of the forest. All, however, are unanimous in recognizing that the region is the last and the most important repository of tropical fruit species. Many are still unknown to botanists, agronomists and the general public as they occur in inaccessible places in the great forest. They are, however, well known to Indian and caboclo *populations, who rely on them as an important complement to their diet. Some fruit, even more timid, hide even farther away, sharing their mysteries only with the wild animals, awaiting the moment to cry, from their sea of chlorophyll, "here we are".*

It is not just the extraordinary diversity of fruit trees that holds one's attention, but also the great variation existing within each species. Plants of the same species flourishing splendorously in a multiplicity of shapes, materializing the miracles of mutation, segregation and genetic combinations. Such is the case, for example, with the abiu, in which the weight of the fruits in some trees is a mere 30 g, while in others it may exceed 1 kg; and the pupunha, some with thorns and some without, and whose fruit may be large or small, oily or floury.

This abundant source of flavors, aromas and vitamins carries within it the secrets of the forest. Almost all the species have strange names but which, etymologically, always bear some important significance, that the simple language of the forest folk manages to condense into a single word: bacuri *is a fruit that falls as soon as it ripens;* abiu *is a fruit with a soft skin;* açaí *is the fruit that weeps water;* cupuaçú *is the one that looks like a cacao fruit, only larger.*

But what can one say of all these fruits? What to say, for instance, of the açaí, a small fruit which, when transformed into a liquid, is a part of the daily food for a large portion of the Amazonian population, shared by rich and poor, by those on river banks or on asphalt streets, in a veritable gastronomic ecumenism.

And the cupuaçú? This is a very large fruit, with a thick skin which is unable to hide the distinctive and unmistakable aroma of its delicious pulp. The cupuaçú is gradually winning enough status to justify its cultivation and become a well-known plant. There are already ten thousand hectares under cultivation in Amazonia! That is a lot for a plant that not so long ago kept itself hidden under the shade of bigger trees. And it is undergoing changes imposed by this domestication process - its height under cultivation is not so great, its productivity is beginning to increase, and it does not require so much shade to be productive.

The Brazil nut tree is another example of the forest's opulence. The most convincing theory is that the extensive native Amazonian nut tree groves would have been pre-Columbian Indian cultivations, later abandoned and incorporated back into the forest. Exploited since the 19th century under an extractive system, only in recent years has the nut tree - a veritable vegetable monument - been studied with intensity, which has made it possible to begin cultivating it.

The species has very little natural regeneration, and, in spite of being protected by Federal Laws against logging, the tree is being seriously impaired by the degradation of the environmental conditions in the region. Burning the forest has decimated whole populations of the trees or else rendered them unproductive. It is the silent protest of a plant that, without insects to pollinate it, simply ceases to bear fruit.

All these species together with the araticum, bacaba, buriti, murici, bacuri, sapota-do-solimões, sapucaia, araçá-boi. araçá-pêra, biribá, sorva, pequiá, mucajá and many many others, are triumphantly preparing for the "era of new cultures" when they will no longer be a mere object of curiosity in botanical gardens, but important components in the world's tropical orchards. We eagerly await this new era, while Silvestre Silva's photographs and Helena Tassara's text so ably describe the beauty and the flavors of Amazonian fruit.

José Edmar Urano de Carvalho and Carlos Hans Müller
(EMBRAPA - CPATU)

NATIVE FRUIT OF THE SAVANNA

The region encompassing the "Cerrado" or Savanna ecosystem reaches mainly into the states of Minas Gerais, Goiás, Mato Grosso, Mato Grosso do Sul, Tocantins, Bahia, Maranhão, Piauí and the Federal District, and corresponds to approximately 25% of the Brazilian territory.

Due to its size and geographic distribution, this region presents a great variety of climates, soils, fauna and flora.

The climate usually consists of a rainy season (spring and summer) and a dry season (autumn and winter), which are well defined.

The soils are old and mostly deep, well drained, with a flat or gently rolling surface, high acidity, low natural fertility and a poor capacity for storing water.

Despite the limitations imposed upon the growth and development of the plants, due to rainfall and soil characteristics, the Cerrado ecosystem is surprising for its enormous variety of plant species. The area contains over 40 physiographical types of landscape, among them the cerrado *(savanna),* cerradão *(barren land),* campo limpo *(grass land),* campo sujo *(scrub land),* vereda *(bottom land),* mata de galeria *(riverside forest) and* mata calcárea *(calcareous forest).*

This vegetation, still largely unknown, has a great potential for food, timber, fuel, farming, fodder, medicine and ornamentals.

As of the 1960's, with the Federal Capital being transferred to Brasília, right in the heart of the Cerrado, with the construction of highways and the adoption of a policy of development of the interior and national integration, this region suddenly became important for the production of food and energy. Thus, from a small agricultural subsistence activity heavily dependent upon cattle-raising, the region began to contribute a large proportion of grain production and to contain a significant share of the country's cattle herds.

In spite of the existence of environmental protection laws, farmers, mainly coming from other regions, have indiscriminately destroyed the native vegetation due to their ignorance of its potentialities.

Within this scenario, the Cerrado ecosystem has been attacked and destroyed by fire and tractors, placing various species of plants on the brink of extinction, including some native fruit trees before they have even had a chance to become known.

Therefore, to show the potential for making use of these plants is the only way to save them and promote balanced and sustainable farm development, ensuring a better quality of life for future generations.

In the Cerrado ecosystem there are countless native fruit trees with a fundamental importance for both animal and vegetable wild life and for the maintenance of this biome.

The consumption of native Cerrado fruits, for thousands of years practiced by the Indians, was of fundamental importance for the survival of the first explorers and colonizers of the region. Adapting and creating techniques for processing these fruits, man invented veritable treasures of today's culinary art, such as liqueurs, sweets, jellies, cakes, beverages, ices and appetizers. And interest in these fruits has reached several segments of our society: farmers, industrialists, housewives, businessmen, research and technical assistance institutions, cooperatives, universities, health and food organizations.

Industrial interest in Cerrado fruits intensified after the 1940's. The mangaba tree, for example, was intensively exploited during the Second World War for its latex. The babaçu and macaúba were studied closely during the oil crisis in the 1970's and showed great promise for use in internal combustion engines instead of diesel oil. The pequi is already being industrialized, its oil canned and marketed. Macaúba pulp and oil are today used in the manufacture of coconut soap. The bitter-tasting heart of the guariroba shoot has recently been marketed as a preserve, as is the case of the sweet cabbage palm. Baru kernels are already being exported. Ices prepared from araticum (custard apple), pequi, cagaita and mangaba are all the rage in ice-cream parlors in The Federal District and in Belo Horizonte.

Currently, in many regional market places and roadsides, it is possible to find a great variety of fruit, native to the Cerrado, being sold at reasonable prices and attaining great popular acceptance. Today the existence may be noted of a potential and emerging market for fruits native to the Cerrado, a market still to be exploited by farmers, as up to the present all the harvesting of these fruits has been done in a predatory, extractivist manner.

EMBRAPA and other research institutions, through the Cerrados Farm Research Center (CPAC), have been concerned with this fact, and have conducted many studies on seed germination, the production of seedlings, planting, nutritional value, processing and the storage of Cerrado fruit.

Notable among current possibilities for the utilization of Cerrado fruit trees are: planting in environmentally protected areas; enrichment of the flora in poorer areas; recuperation of areas which have been deforested or otherwise degraded; planting in reforestation areas; planting in public parks and gardens; planting in uneven, broken ground; and the formation of domestic or commercial orchards.

Many owners of farms and small holdings are already starting up orchards of fruit trees native to the Cerrado and nursery gardens are stepping up the production of seedlings.

From a nutritional aspect, native Cerrado fruits have high contents of sugars, proteins, vitamins and mineral salts. Another great virtue of these Cerrado fruits are their singular flavors, able to excite the palate of anyone, capturing his appreciation.

The fruits native to the Cerrado ecosystem presented in this book represent the species most utilized by local communities, and are also those with the greatest economic potential. It is to be hoped that one day these species will come to be planted on a commercial scale, thereby saving them from extinction, increasing farmers' incomes, providing raw materials for agribusiness and healthy food for the population.

Djalma Barbosa da Silva
José Antonio da Silva

(EMBRAPA - CPAC)

Savannas, forests, meadows, tablelands, jungles, scrub lands, coast and hinterland: everywhere in this continent-country a fabulous variety of plants live in permanent transformation, adapting to the most diverse conditions of climate and soil, to be able to grow, flower and bear fruit.

Right along with them, man, or rather the Brazilian man, with his marvelous and genial cultural diversity, has over the centuries learned to take advantage of all this exuberance. Sometimes these same men have overdone their exploratory eagerness and caused veritable ecological catastrophes, but in general they continue to enjoy the pleasures and benefits these fruits have to offer to those who know how to - and wish to - enjoy them.

Without a shadow of a doubt, the land of Brazil, Pindorama, a fertile and abundant land like very few other regions on the planet – possibly like no other! – harbors an impressive natural orchard while still retaining the capacity to receive and cultivate new plants.

This fact was known to the native Brazilian since the days of his forefathers and did not escape the notice of the European colonizer. It was a hope of survival for the slaves transplanted from their African homeland, and has been well made use of by a host of foreigners who have settled here. To all these men, who over the years have together formed the Brazilian nation goes the privilege of continuing to enjoy these fruits. A privilege which has now been spread around the world with the publication of this beautiful work by Empresa das Artes.

In preparing the text presented in this book, I was guided by the beautiful and unique photographs of Silvestre Silva and his extraordinary passion for fruit. I was also fortunate enough to have been followed closely by the patient and wise dedication of Flávio Carvalho Ferraz, a defender of Brazilian Mirtaceae, his heritage from Minas Gerais. I also had the precise and timely scientific guidance of botanist Sérgio Romaniuc Neto.

As much as possible, I began to fill it all in with the colors, aromas and flavors of my own discoveries, a precious fund of information that, little by little, I found in libraries, encyclopedias, manuals, books, magazines, and especially in the many hours of agreeable conversations with friends and enthusiasts on the subject.

To write about all these fruits, the fruit of many plants, was for me an enormous, sensitive and surprising apprenticeship, which I hope to share with the readers.

Helena Tassara

ASSAI

Scientific Name: *Euterpe oleracea* Mart.
Botanical Family: Palmae

Origin: Brazil - Lowlands and river banks in the Amazon region.

Plant Features: A palm tree with a thin, elegant trunk, may rise to 25 m in height. Large leaves, divided into thin dark-green strips, frequently reaching 2m in length. Small yellowish flowers grouped in large pendulous bunches, flowering predominantly from September to January, but may appear all the year round. A species very similar to *hearts-of-palm* (*Euterpe edulis* Mart.) of the Atlantic forest, different in that it grows in groves of from 3 to 25 palm trees.

Fruit: The fruit grows in bunches and is violet in color, almost black when ripe. Round or egg-shaped, it contains a rich edible pulp and a hard seed. Produced during most of the year, but more abundantly during the months of July to December.

Cultivation: May be grown from seed or from shoots at the base of the tree. Flourishes well in various types of soil and climate, preferably in hot regions. Each palm tree produces 3 to 4 bunches of fruit per year; each bunch having from 3 to 6 kg of fruit.

The Assai, a typical plant in the Brazilian tropics is, along with the Buriti, a fundamental part of the Amazonian forest landscape. The Assai grows well as much on dry land as in bottom lands subject to periodical flooding, as long as the water is being constantly renewed. A perennial, riverside plant, the Assai is also important for protecting the soil during heavy tropical rainfall conditions.

Everything the Assai produces is useful. Fruit, fronds, roots, the hearts-of-palm, trunk and fruit stalks. From Santarém to Marajó Island, the Amazonian riverside population know they can count on this tall, elegant palm to nourish and sustain their families practically throughout the entire year.

Assai fronds are used for roofing houses, the fibers for weaving hats, mats, handbags and rasas - baskets used as standard measures in extractive activities practically everywhere in Amazonia. The trunks when dry provide a highly durable timber, unusually resistant to insects and decay, which is widely used for building houses, bridges and other constructions. Even the dried stalks, after the fruit have been removed, may be used as brooms.

The Assai is also a generous source of popular medicine: the unripe fruit is used to cure intestinal problems; the roots used as a vermifuge; the palm hearts in paste form act as an anti-hemorrhagic agent, frequently used after extracting teeth.

But above all the Assai palm gives us the little, purple Assai fruit, almost black when ripe. It is a small coconut from which is extracted, through crushing, the traditional and popular "Assai wine or juice".

Assai juice has a large market throughout the Amazon region, reaching the astonishing estimated figure of 100 thousand to 180 thousand liters/day in the city of Belem alone.

Although the tree produces fruit practically all the year round, the native knows that when the mother of the sun - the cricket - sings in early summer to announce the beginning of the dry season, this is the beginning of the best Assai harvest. He it is who will climb the 20-meter high palm trees to reap the fruit. With a great deal of skill, strength and balance, the collector climbs the trunk until he reaches the loaded bunches, aided only by a peçonha - a kind of strap made of the green leaves of the Assai itself - looped around his ankles to assist his climb. Once at the top, having cut down with his machete only the ripe bunches - in their ideal state they are called tuirá, when they are black with a whitish cast to the skin - the native then begins to swing the slender palm like an inverted pendulum, until he can reach the neighboring palm and move to that one. Right there or on the boat waiting at the bank of the backwater, the bunches are pulled apart and the fruit stored.

However, the Assai has to be processed daily due to its rapid fermentation. It will not last more than 24 hours even when refrigerated, so it is that every day, 7 days a week, before the day has fully broken, the sidewalks around the Ver-o-Peso Market in Belem are cluttered with baskets, sacks and tin cans filled with the little purple fruit arriving on trucks and in boats from the interior of the state.

Once it has been sold in the market, the Assai will be turned into juice. The fruits are first placed in warm water to soften the pulp and loosen the stones and then crushed by hand or in a special machine. Then, as Câmara Cascudo so aptly describes it, "from the blood-red, purplish paste, run through a sieve if hand crushed, dissolved in several changes of water, the wine is formed - the Assai drink, thickened to the taste of whoever is to drink it".

While it is commercially processed, many families in Pará produce their own Assai juice for daily use, selling the remainder from their homes, advertising the sale by means of a small red flag placed at their doorways.

A tradition probably inherited from Amazon indian groups, Assai juice today takes a basic role in the nutrition of the regional population. It is consumed pure at any time of the day in the form of beverages, juices and ices, with or without sugar, in the morning instead of milk, and is given even to small children at all their meals, thickened with manioc flour or tapioca. It may be served along with fish or dried shrimp, meats or beans and rice, or even pure when there's nothing else.

Today it is known that Assai juice is essentially an energy-producing food, with a high calorie count, with 2.37% of protein and 5.96% of fat. In spite of its undeniable importance in commerce, in the daily diet of everyday life, the Assai tree is today the main source of hearts-of-palm in the country. Close to 95% of all the local production of palm cabbage comes from the Assai palms in Amazonia, as the native reserves in the Atlantic Forests are virtually depleted, where the palms have been decimated with no concern for preservation.

In the current extractive system related to the Assai palm, the fruits are used for local consumption and the palm cabbage for export. Organized, rational plantations are still rare, and the destruction of Assai palms still continues in a systematical, illegal, unbridled and clandestine manner. Few have been concerned with its preservation, but the results obtained from experiments conducted by EMBRAPA technicians with the planting and management of the Assai have begun to awaken the interest of farmers and businessmen, especially attracted by the potential prospects of the product on the domestic and foreign market.

As a plant, one of its main characteristics is the fact that it grows in clumps which, if allowed to develop, may amount to as many as 20 trees with varying ages and degrees of development. It is exactly the good management and rational culling of the plants and new shoots, as well as the selection of those trees most suitable for bearing fruit, for extracting palm cabbage and for gathering seed that makes its profitable commercial exploitation possible.

This is a great challenge for the forthcoming years: to exploit the Assai fruit in a rational manner, so as to preserve its resources for the future.

AVOCADO

Scientific Name: *Persea americana* Mill.
Botanical Family: Lauraceae

Origin: Central America.
Plant Features: A tree with a brownish bark, may reach 20m in height. Foliage always green. Small greenish-white flowers with a characteristic strong odor attractive to bees.
Fruit: The fruit has an olive-green shiny skin and a thick, creamy pulp-colored creamy yellow, rich in vegetable oils. The hard seed is surrounded by the soft pulp.
Cultivation: Can be planted anywhere in Brazil, preferably inside orchards to avoid strong sunlight reaching the trunk. Should be planted preferably at the beginning of the rainy season in permeable, deep well fertilized soil. Reproduction most indicated is through grafting.

In America it is rare to find a back yard, an orchard or spare plot of land in homes, holiday homes, or farms that do not have at least one avocado tree bearing fruit. In rural areas and in cities, most of the population know and make use of the avocado tree and its fruit.

Originally from tropical America, in the regions colonized by the Spanish (Mexico, Guatemala and the Antilles), the avocado spread to South America and Amazonia, and today is to be found anywhere in the world where there is fertile soil and a sufficiently warm climate. Avocado producers and exporters are to be found in all the countries of Africa, South and Central America, as well as Israel, Spain and the United States, in the California region.

According to Pio Corrêa the avocado was introduced into Brazil as a cultivatable species only in the early 19th century. Today it is on sale in markets and fruit stalls practically all year round. In spite of being the world's largest producer of the fruit, Brazil exports less than she could because, ironically, her avocados, large and attractive, are strange-looking to European consumers, used to smaller fruit. Plantations in the interior of São Paulo State, especially in the region of Campinas and in the state of Minas Gerais are responsible for almost two thirds of the total production, followed by crops from the states of Paraná, Espírito Santo, Pará and Rio Grande do Sul.

With a mild, tempting flavor, not too sweet or bitter, the avocado is a smooth, generous and meaty fruit. Although there are various varieties under cultivation, presenting large variations in the shape, size, weight and coloring of the fruit, with different types of pulp, fiber, stone and skin, in Brazil, according to Eurico Teixeira, the avocado fruit has always been one: simply an avocado.

The name by which the fruit became known may have been a derivation of the Maya word *ahuacacuahuitl*, or the Aztec name *Ahuacatl*, both sonorously transformed into *abogado* or *aguacate* in Spanish and *abacate* in Portuguese.

Some say that the Maya name signifies "fruit with the shape of a testicle or scrotum", the inference being that the avocado would have aphrodisiac properties. Others say that the Aztec name signifies "butter coming from wood", in a precise description of the sensitive properties of the fruit. It is a fact that the creamy greenish yellow or pale yellow pulp, strongly resembling a buttery cream, is basically comprised of non-saturated fatty acids, containing only 70% of water in its composition, which is little as compared to the majority of existing fruit.

For Pimentel Gomes, the avocado tree is one of the most valuable fruit trees in existence, as its fruit contains large percentages of fats, carbohydrates and proteins, together with a wide range of vitamins. The avocado is an incomparable source of energy, with a high food value and a low cholesterol content, which, contrary to what used to be popularly believed, renders it easily digestible with no side effects.

Due to its qualities and to its extreme smoothness on the palate, the avocado is one of the most versatile fruits in axistence, a source of inspiration for countless and varied recipes. Due to the mild flavor of its unsweetened pulp, the avocado may be eaten either as a sweet or together with salty foods, according to the habits and cultures of the people in the regions in which it is cultivated, often shocking one another. For the Brazilian Eurico Teixeira, for example, the avocado eaten as a salad "seems as if it must be horrid".

This disparity in tastes is explainable: in Brazil the fruit *in natura* is eaten only in its sweet variations, as deserts or snacks. The avocado pulp in pieces, mashed or creamed may be mixed with sugar or honey, with or without a few drops of lemon juice; it may also be blended with milk or the juice of other fruits and drunk as a *"vitamina"*.

In Europe, Central America and in Mexico the avocado is used as a essential ingredient in salads. Also *in natura*, in pieces or creamed, mixed with greens, tomatoes, onions, aromatic herbs and seasoned with mayonaise, olive oil, salt, pepper or other spices, the avocado is served with toast, bread or shrimp in salty dishes.

In Guatemala, Porto Rico, Mexico and Colombia, it is the custom to garnish soups with thick slices of avocado. In some places in the Carribean, avocado is eaten with Port or Madeira wine, with orange or lemon juice. In most Latin American countries the famous and international *guacamole* is prepared and served with corn *tortillas*.

If it is true that there's no accounting for taste, here's one good reason to try new recipes and start the discussion!

"The Araçá trees are another species that grow mainly in poor land close to the sea, resembling apple trees in size, color of the bark, the scent of the leaves and in general appearance. The flowers are white and sweet-smelling. The Araçá fruit is about the size of a medlar or even larger. When unripe it is green in color, and when ripe the color of a pear. The fruit is eaten whole, and has a very tasty tang to it, it is also used to make a delicious sweet, which is very good to cure the cramps".

NOTÍCIA DO BRASIL
Gabriel Soares de Sousa (1587)

ARAÇÁ

Popular Name:
Scientific Name: *Psidium* sp.
Botanical Family: Myrtaceae

Origin: Brazil - Amazon region, and from the Guianas down to the State of São Paulo.
Plant Features: Trees may vary from 3 to 10m in height, with a smooth, scaly bark and sparse foliage. The flowers are usually reddish when young. Greenish-white flowers.
Fruit: Spherical, green, yellow or red fruit, depending upon the species. The yellowish-white or reddish mucilaginous aromatic pulp contains many seeds.
Cultivation: Found in Brazil in a wild state, it prefers dry soils and is not particular about the climate, even being resistant to frost. Propagates from seed and grafting.

Of all the many types of araçá in Brazil, to which was the traveller referring? Was he actually referring to one of the araçás or could it have been a guava? Or, to increase the confusion, could it have been an araçá-guava, as the guava is to this day still called in some regions of Bahia? Difficult to tell! The fact is that araçás and guavas in the wild are very similar and both belong to the same family of Myrtaceae.

The araçá is usually far more acid than the guava. Like the guava, it also has a soft pulp full of seeds, but the majority of varieties are less meaty and therefore economically less valuable.

In addition, there are almost as many types of araçá as there are beaches in Brazil: white araçá, gray araçá, pink araçá, red araçá, green araçá, yellow araçá, wild araçá, beach araçá, field araçá, party araçá, araçá-de-Minas, araçá-de-Pernambuco, araçá-do-Pará, araçá-de-corôa, araçá-boi, pear araçá, butter araçá, araçá-de-folha-grande, araçá-de-flor-grande, small araçá, araçá-mirim, araçá-guaçu, araçá-peba, araçá-pitanga, araçaí, araçanduba, common araçá, true araçá, or simply, araçá.

These many varieties of araçá are to be found growing all over Brazil, from the Southern fields to the Amazon jungle, preferably where it is hot and damp. When referring to species typical of the Amazonian region, Paulo Cavalcante states that the araçá is a plant of many varieties, dispersed all over the country, and may be found *"either cultivated or wild, in open fields or areas of thin, low vegetation varying greatly in size, from a small 70 cm high shrub to a small tree from four to six meters in height.*

Some species of araçá bear fruit that is very tasty and are best appreciated when eaten directly beneath the tree when in season. Others have fruit that is too astringent and acid, and can only be used for making preserves, where their sharp flavor imparts a special sweet-sour flavor. Notable among the preserves produced is an araçá paste, called *marmelada de araçá*, which is very similar to guava preserves and pastes.

As with most fruit, trees and plants, the araçá and its leaves are widely used in popular medicine.

One species of araçá that stands out is the araçá-boi (*Eugenia stipitata* Mc Vaugh). Although it is native to Peruvian Amazonia, where it is well-known and widely used by the regional population, it also grows in the state of Acre in Brazil. The araçá-boi is a small tree, almost a bush, never taller than 3 meters in height. The canary-yellow fruit varies in size, but is always large, larger even than the largest cultivated guavas, and may weigh up to 400 grams of pure pulp. Hence its name of *araçá-boi* (bull araçá). The pulp is juicy and, although quite acid, is tasty and, for this reason, is consumed more in the form of ices, sweets and beverages than *in natura.*

Psidium humile Vell

BARBADOS CHERRY

Scientific Name: *Malpighia glabra* L.
Botanical Family: Malpigiaceae

Origin: The Antilles - Central America.
Plant Features: A bush, up to 3m in height. Trunk branching from the base. Dense foliage of small, dark green shiny leaves. Flowers disposed in bunches, colored pale rose or violet. Flowers all the year round.
Fruit: A pulpy fruit with smooth skin and divided into 3 sections, orange-red color when ripe. Contains three seeds. Fruit appears 3 to 4 weeks after the flowers.
Cultivation: Grows in any type of average fertility, well drained soil. Prefers hotter regions (25ºC to 27ºC). Can be cultivated all the year round, especially in the rainy season. It is slow-growing and may be used as an ornamental plant.

The cherry of the Antilles, or Barbados Cherry, leaves no doubt about its origin or its appearance; it is the name of a very pretty and useful fruit, not unlike its European namesake.

Bright red when ripe, or varying from orange to purple tones, with a perfume similar to that of an apple, a slightly acid flavor and a soft, juicy pulp, the Barbados Cherry has been eaten for centuries by the natives of the Antilles, Central America and Northern South America.

A hardy, rustic plant, the Barbados Cherry propagates naturally and easily everywhere. It is a medium-sized bush, which develops well in tropical and subtropical climates, or even in semi-arid regions, as long as there is some regular supply of water. It does not have many requirements as concerns the soil. Fruition begins from two to three years after planting, and may occur in abundance between four and seven times a year.

For a long time this tropical "cherry" from the Antilles remained flowering and producing its fruit in the Americas without attracting a great deal of attention. It did not have the glamour of the cashew or the variety of the sweetsops. It wasn't until the 1940's, when scientists in Porto Rico discovered high concentrations of ascorbic acid, or vitamin C in the edible portion of the fruit, that interest was awakened in the Barbados cherry and its economic potential.

The Barbados cherry concentrates, in the same amount of pulp, approximately one hundred times more vitamin C than that in an orange or a lemon, twenty times more than in a guava, and ten times more than that contained in a cashew or a mulberry. Four fruits per day are enough to provide all the daily vitamin C requirements of a healthy adult person.

Today it is known that, due to its ascorbic acid content, the Barbados cherry is not just indicated as a health – maintaining item, but it also prevents debility, irritability and fatigue, loss of appetite, and reduces the occurrence of infectious diseases, muscular and articulation pains. It can also be used to fight colds and pulmonary infections, in controlling cases of persons suffering from tendencies toward nasal and gum hemorrhages and as an aid in treating liver diseases. For this reason, it is also indicated in the diet of nursing mothers, children and adolescents, in expectant mothers and wet-nurses and to strengthen aged, undernourished or debilitated organisms.

Held to be a State Secret, the Barbados cherry remained imprisoned in Porto Rico until it was brought secretly to Brazil in 1956 by a professor of the Federal Rural University of Pernambuco. Of the 245 seeds planted on the University's campus only ten germinated and became productive plants, and it is probable that the majority of the seedlings currently planted in Brazil are descendants of those first original plants.

During the 1980's the UFRPE sponsored and developed an enormous campaign to make people aware of the nutritional values of the Barbados cherry and its potential uses, as well as the methods of cultivation and care required for its growth, which are not many, due to its hardiness and rusticity.

The new little ancient fruit was widely accepted and immediately successful. In some cities in the interior of Brazil, for example, there is practically no back yard or orchard that does not contain a Barbados cherry tree. It was, however, the great agro-industrial groups that quickly appropriated cultivation of the cherry, for the purpose of exporting the frozen pulp and its by-products, and leading Brazil to become the world's largest producer, consumer and exporter of the fruit.

In Pará, in the Brazilian Northeast, and in some locations in the interior of the state of São Paulo many hectares are already planted in Barbados cherry, producing enough to supply the international demand, which is growing rapidly. This is because the Barbados cherry fits perfectly into the worldwide trend, started in the 1980's, of seeking for natural health products.

Half-way through the 1990's, all over the world, the Barbados cherry is being widely consumed, be it *in natura* or in juices processed from the frozen fruit, in jellies, jams, preserves, liqueurs or beverages.

The pulp of the fruit is also widely used to enrich the vitamin content of other fruit, where the ascorbic acid also acts as an anti-oxidant and natural preservative, and a paste made from the unripe fruit is a raw material for producing vitamin pills for those who find the fruit too acid for their taste.

PINEAPPLE

Scientific Name: *Ananas comosus* (L.) Merril.
Botanical Family: Bromiliaceae

Origin: Central America and Mexico.
Plant Features: A small plant, may reach 80cm in height. Long, hard leaves arranged in a spiral from the base, in a rosette. Small flowers, colored pink or mauve-purple, agglomerated on one stem, forming a spike which develops into the pineapple fruit.
Fruit: A group of small fruit form the egg-shaped structure of the pineapple. On its upper portion a crown of hard, intense green-colored leaves form, called bracts. The internal stem of the pineapple is surrounded by a succulent edible pulp.
Cultivation: The plant prefers a nutrient-rich soil, does not accept wetlands. Planting is done at the beginning of the rainy season. It develops well in locations with temperatures from 6ºC to 20ºC, free from frost. Each plant produces 20kg per year.

When Christopher Columbus arrived at the is land of Guadelupe in the New World, the European invaders were offered pineapple in a gesture of hospitality and welcome. For its resemblance to the fruit of the European pine tree, the new fruit was called *piña*, as it is known until today in Spanish-speaking countries.

Indian tribes speaking the Guaraní language had a name that meant "tasty fruit", from which was derived the name *ananás*, and some wild species are still known as such. For indian tribes speaking the language, their name, *iuaka'ti* meaning "fragrant fruit" later gave rise to the Portuguese *abacaxi*.

Probably native to Southern South America, now Paraguay, the pineapple was carried all over America by the Guaraní indians and eventually the species came to be cultivated by indigenous populations as far away as Central America and the Carribean, long before the arrival of the Europeans.

Under the name of *Piña*, the pineapple was taken to Europe as proof of the exotic exuberance of the lands west of the Atlantic. A fruit that is easily dispersed and cultivated, requiring only heat, the pineapple crossed the oceans of the world aboard galleons and caraveles, arriving in Africa, Madagascar, China, India and the Phillipines. In these locations the pineapple propagated easily and rapidly, and has been very well exploited over the last five centuries, and not just as a tasty fruit.

In England it became a veritable sensation, and, in the 17th century, pineapples were cultivated in greenhouses specially built to maintain a temperature equivalent to the tropical climate required for their growth. With its thorny crown it became known as the "queen of fruit". Transformed into a delicacy by kings and queens, the pineapple was served up as a symbol of hospitality to special guests of nobility, also in European courts.

In its passage from the New to the Old World, the pineapple became more than a fruit, as a veritable model of beauty and exoticism, tirelessly represented in paintings, studied and admired by natural scientists. Its image was one of mystery for a long time, until botany completely unveiled its secret. It was eventually discovered that what had been considered to be one single fruit was in truth the fusion of several hundred small fruit into one large body.

With its strong aroma and a flavor that varies from very sweet to very acid, the pineapple has a refreshing, juicy pulp, and is widely used for making preserves, crystalized fruit, jams, juices, ices, creams, jellies and puddings. In Brazil a drink is made, called *aluá*, which is well-known and appreciated in the Northeast, from pineapple skins, steeped in water for a few days, until they ferment.

The pineapple is undoubtedly one of the most popular tropical fruit in the world, and is widely used in the preparation of cocktails with a festive, Carribean spirit, such as the world-famous *piña colada*, made from pineapple juice and rum. It does not contain many calories, but it does have a large amount of vitamins A, B and C, as well as carbohydrates, mineral salts and fibers. Also from the remains of the pineapple bromelin may be extracted, a valuable enzyme used to digest proteins, leaving a consistent bagasse that can be used as animal feed.

Brazil is one of the world's largest producers of pineapple, as of the early 1990's, losing her leadership only to Thailand, the Phillipines and China. The main Brazilian plantations are to be found in the Northeastern region of the country, in the states of Paraíba, Bahia and Rio Grande do Norte, in the Minas Gerais Triangle, in Espírito Santo and São Paulo, where the municipalities of Araçatuba and Bauru are the leaders.

Although the country has a larger area under cultivation than any other producing country, Brazil still does not have the technology to reach the high levels of productivity obtained in Asia. Basically, Brazil cultivates varieties such as **pérola** (pearl) preferred by the *in natura* market, and **smooth cayenne** or **havaiana** (Hawaiian), which is a larger fruit, more acid and sturdy and, for this reason, usually destined for export and for the preserves and juices industries, which usually have their own plantations.

AROEIRA-VERMELHA

Scientific Name: *Schinus terebenthifolius* Raddi
Botanical Family: Anacardiaceae

Origin: Various Brazilian formations.
Plant Features: A tree 4 to 10 m in trunk height with thick bark and dense foliage. Leaves composed of 3 to 10 pairs of leaflets with serrated edges. Small, pale yellow aromatic flowers that attract bees appear between September and January.

Fruit: Brilliant red, rounded fruit in pendulous bunches, having a reddish, edible pulp, eagerly sought by birds.
In season from January to July.
Cultivation: A wild plant, rarely cultivated. Also used as an ornamental.

The small, very ornamental aroeira tree grows over much of Brazil, from the state of Pernambuco down to Rio Grande do Sul.

Many plants are known as aroeiras in Brazil. By their popular names, it can be seen that there are species scattered all over the country. Besides the aroeira-vermelha (red aroeira), there are: aroeira-da-praia, aroeira-de-bugre, aroeira-do-sertão, aroeira-do-mato, aroeira-da-capoeira, aroeira-do-campo, aroeira-de-goiás, aroeira-do-amazonas, aroeira-do-rio grande, aroeira-rasteira, aroeira-mansa, aroeira-brava, aroeira-preta and aroeira-branca.

Specifically, the aroeira-vermelha, a typical plant in the Northeastern Brazilian scrublands, is native to Brazil. According to Luiz Emygidio de Mello Filho, together with other trees, such as *barriguda-do-sertão*, *baraúna*, cashew, *juazeiro oiticica and umbu*, the aroeira is very important throughout the region.

The small, bright red shiny fruit are sought after avidly by birds. When in fruit, the tree becomes very attractive, as it is covered all over with clusters of fruit on the ends of the branches. In some places, it is believed that the essential oil contained in the plant is poisonous, but in actual fact what occurrs is no more than an allergic reaction for some sensitive people. For this reason, the aroeira-vermelha is also *falsa-pimenta* (false pepper) or *aroeira-mansa* (tame aroeira).

The small, round shiny seed of the aroeira-vermelha is used as a spice, to add flavor and refinement to dishes all over the world. The gentle, slightly peppery flavor of the aroeira-vermelha, as well as its attractively decorative appearance, call for its use in many dishes, either as whole grains or ground up. It is especially used to season sauces for white meats such as birds or fish, as is not too overpowering for these subtle flavors. Introduced into European cooking under the name of *poivre rose* (rose pepper), the aroeira-vermelha adds a tropical touch to *nouvelle cuisine.*

From the skin of aroeira seeds, especially from the aroeira-vermelha, an oil can be extracted, which is often used in popular and rustic medicine, reputedly effective for treating fevers, tumours, diseases of the cornea and the respiratory and urinary tracts. Seeing is believing!

Annona crassiflora Mart.

SWEETSOP

Scientific Name: *Annona* sp

Botanical Family: Annonaceae

Origin: The Antilles.

Plant Features: A variable-sized tree, may reach 7m in height, depending on the species. Rigid leaves arranged characteristically, interspersed and horizontal along the branches. Flowers are frequently fleshy, pale green or yellowish-white, appearing all the year round.

Fruit: Globular or elongated containing numerous seeds surrounded by a soft, white aqueous pulp and a greenish yellow smooth or scaly rind. Fruition occurs almost all the year round

Cultivation: Propagated from the seeds or cleft grafting. The plant prefers a hot climate but with little rain and a clearly-defined dry season. It begins to produce 3 years after planting.

The *Annonaceae* family is comprised of a large variety of exotic fruits, usually delicious. In general, the plants of the custard-apple family are characterized by having simple leaves disposed alternately on one same plane along the branches and by the similarity between the fruits.

The fruits are well-known and much appreciated all over the world, with their rustic appearance, reminding one somewhat of a pine-cone; in shape they may be elongated or globular though never completely so; sometimes, depending upon the species, they may be heart-shaped. Size and weight also vary, from the size of a fist, to as large as a rugby ball. In Brazil the names also vary, such as Fruta-do-conde (Sweetsop), (*Annona squamosa*), Graviola (Soursop), (*Annona muricata*), Araticum-do-serrado or Marolo (*Annona crassiflora*) and many others.

According to Maria do Carmo C. Sanchotene, the Guaraní Indian name *araticum* means "soft fruit", which the rough-looking fruit becomes when ripe, disintegrating easily.

Araticum is in fact the most common name for the wild varieties of *Annonaceae* throughout the American continent where Portuguese is spoken. In Spanish-speaking America, the generic and more common name for the fruit is *anone* or *anona.*

Colombian writer Clara Inés Olaya provides a clue toward understanding the motives leading to these generalizations. According to her, the Spaniards would have tasted these fruits in the Caribbean Islands, when they first arrived in the New World, as they are native to that region. There also they would have learned the name *anón*, from the Taína Indian-language word used to designate a given variety of native fruit, which later became known as *"the Spaniards' custard"*.

The *anon* described by the 16th-century navigators grows on shrubs, is dark green, with a scaly, rough texture and, when ripe, contrary to what its appearance would suggest, is easily broken into segments with the fingers. Once open, the fruit offers a soft, creamy pulp that melts in the mouth, leaving a pleasant, perfumed flavor, and contains small, hard, shiny seeds.

This description is precisely that of the *Annona squamosa* or sweetsop, among all the *Annonaceae* the best-known and most commonly to be found in open-

Annona squamosa L.

air markets and supermarkets all over Brazil, although it is not native to the country.

In fact, the fruit's introduction into Brazil has exact historical dates: according to Pio Corrêa, the Count of Miranda in 1626 planted the first tree of this variety in the State of Bahia, but it was not until 1811 that a French agronomist introduced it into Rio de Janeiro at the request of the King D.João VI. It is thought, however, that some wild varieties of the fruit, originally from the Antilles, may have migrated into the Amazon region, becoming transformed into sub-spontaneous species before the arrival of the Europeans.

Currently in Brazil, due to so many moves from here to there, the Spanish *anón* has a variety of names in Portuguese: it may be ata in the North and Northeast of the country, or in the states of São Paulo and Minas Gerais; it may be *araticum* in Rio Grande do Sul, while in Bahia the fruit may be known as Fruta-do-conde (the Count's fruit) or as *pinha.*

Whether it's an ata, pinha or Fruta-do-conde, the sweetsop is currently cultivated throughout the country. With its delicious, typical flavor, the sweetsop is usually eaten *in natura*, but its pulp may also be used for preparing juices and other beverages.

In the 16th century, as the Europeans advanced into the new continent, they found other varieties of similar fruit. Very like one another, according to Clara Inés Olaya, these fruits were all named by the Conquistadors as *anón*, regardless of the different names the Indians had for them. It is quite easy to imagine that the same occurred with the various Brazilian araticuns.

Among all the known fruits on our planet, none has as complicated a family as the *Anonaceae*, due to the many existing varieties, the resemblance between one fruit and another and the many popular names they have been given in the course of time and history.

The araticum-do-cerrado (Annona crassiflora) is another of them. It is known under this name in the central region of Brazil, in the cerrados (semi-arid scrub lands), from where it derived its name. It is called a Marolo in the south of Minas Gerais, where it is native and spontaneous to the enclosed fields of the region.

Basically, as regards the quality of the pulp, there are two types of fruit: the araticum with a pink, sweeter and more tender pulp, and that with a yellowish, harder and slightly acid pulp. In either case, the process for extracting the pulp, which can be frozen, is similar, being slow, manual and unprofitable.

Whatever name it goes under, the large-sized fruit is well-known and widely consumed by local populations, and is sold in markets and along roadsides when it is in season.

Among native Brazilian fruits that have not become transformed into cultivated species, the araticum-do-cerrado is the one with the most culinary uses. Besides being consumed *in natura*, there are countless recipes for sweets and drinks with the strong, perfumed flavor of its pulp, often enhanced with the flavors of other fruit: in cocktails, liqueurs, soft drinks, cookies, cakes, ices, creams, jams, gelatins, preserves, and other regional sweets.

Annona crassiflora Mart.

ABIU

Scientific Name: *Lucuma caimito* (Ruiz & Pav.) Roem. & Schult.
Botanical Family: Sapotaceae

Origin: Brazil - Amazon region.
Plant Features: A tree up to 10m in height, trunk with rough bark, dense, well-branched foliage. Smooth shiny leaves. Flowers colored reddish yellow.
Fruit: Egg-shaped or spherical, yellow, smooth skin, presenting a milky latex, which coagulates in contact with the air. The pulp is translucent, white or yellowish, mucilaginous and sweet, containing 1 to 4 smooth black seeds.
Cultivation: Grows spontaneously in Amazonia and adapts to cultivation along the seaboard from the far North down to Santos. Cultivated almost everywhere in the country, it prefers deep, humid soils. Reproduces from seed, producing 200 fruits per tree, but may reach a total of 1,000 fruits.

The Abiu is a tropical plant, originally from the Amazon region close to the Peruvian Andes foothills and Brazil's Western Amazon area.

The tree is commonly found wild throughout Amazonia, and some Abiu trees form part of urban landscaping, decorating streets and squares in Manaus, and may even be found in the neighborhood of Belém.

While the tree is best known in Amazonia, it will still grow and bear fruit almost anywhere along Brazil's coastal area, from Oiapoque to Santos, where it has propagated freely.

The shapes of the fruits vary significantly from one variety to another, some being completely round, others oval or even elongated, all approximately of the size of a large chicken's or duck's egg. The skin is smooth and encloses a gelatinous white or yellowish pulp, which is sometimes faintly sweet, at other times tasteless, and sometimes, to everyone's delight, very sweet indeed.

For Eurico Teixeira in his book "*Frutas do Brasil*" (Brazilian Fruit) the Abiu, besides being delicious, is often transformed into a veritable symbol of the country as it bears, like a flag, its main colors: green and yellow. When ripe, some varieties present below the yellow skin a number of green stripes, traversing the fruit longitudinally, from one end to the other.

The fruit is almost always consumed *in natura*, but it may be preserved under refrigeration for up to one week, or processed into a jam.

As a fresh fruit, it should be eaten only when completely ripe and yellow, as otherwise the skin will liberate a viscous form of white latex which adheres to the lips, causing a most unpleasant sensation. This is, in fact, one of the main arguments used to disparage the fruit. On the other hand, this same latex and another, which is obtained from the bark of the tree, are used in the production of an adhesive and for home remedies.

Cultivation of the Abiu tree is apparently simple, as it requires little care, except when it is still young, and little fertility in the soil. Fruition begins at three years of age, ard increases significantly after the fifth year.

In spite of this, the Abiu still remains, since its introduction into the country, as a fruit tree for gardens or non-commercial orchards. As the Abiu is a generous, handsome tree, bearing abundant fruit, despite its limited size, one tree in someone's back yard is enough to supply the entire family, friends and neighbors with the delightful flavor of a succulent, much-appreciated fruit.

AMORA-DO-MATO

Scientific Name: *Rubus rosaefolius* Smith
Botanical Family: Rosaceae

Origin: Southeastern and Southern regions of Brazil.
Plant Features: A fragile, well-branched bush, up to 2m in height, with thorns on branches and leaves. Leaves composed of 3 to 7 leaflets with serrated edges. Greenish-white flowers gathered in clusters at the tips of the branches.
Fruit: In bunches that may reach 20 cm in length. Greenish in color, changing from red to dark purple when ripe. Has an edible, watery pulp.
Cultivation: Grows freely alongside highways and may be cultivated in fertile ground, needing no fertilizer.

Brazil's "wild mulberry" is a little fruit, known almost everywhere in the country, found especially in pastures and along roadsides throughout the southeast and mid-west, preferably at higher, cooler altitudes.

In spite of its name, the fruit, while similar in shape to the real European and North American mulberries, is not related to them. Our wild mulberry is native to this country and belongs to the *Rosaceae* family, along with roses, strawberries, apples and many other delicious fruits and perfumed flowers.

Wild mulberries grow on bushes, frequently in large spontaneous groups. During the season, the wild mulberry grows abundantly and those with the courage to face its many thorns may gather whole buckets of the fruit.

It is appreciated most in its natural state, picked from the bush and eaten immediately, it has a sweet but slightly acid flavor. When, however, the harvest is abundant, providing larger quantities, the fruit may be used to make excellent jams and preserves.

MULBERRY

Scientific Name: *Morus nigra* L.
Botanical Family: Moraceae

Origin: Asia.
Plant Features: A medium-sized tree reaching roughly 4 to 5 meters in height, with a dark, slightly rough bark and good foliage. Light-green leaves, slightly hairy, rough to the feel. Small agglomerated yellowish-white flowers originate the mulberry fruit.
Fruit: Pendulous, dark red, almost black when ripe, with an edible dark-red pulp
Cultivation: Grows well all over Brazil. Fast-growing, it adapts to any types of soil, preferring those that are deep and damp.

Mulberries belong to the *Moraceae* family, which groups together widely varied species, with uses, appearances and flavors so different from one another, such as figs (*Ficus*), bread-fruit and jack fruit (*Artocarpus*), besides the mulberries (*Morus*) themselves.

There are three types of mulberry trees; the red (*Morus rubra*), white (*Morus alba*) and black (*Morus nigra*). The color of the fruit varies with the species they belong to and depends on the degree of ripeness.

Originally from Asia, mulberries were probably introduced into Europe around the 17th century. In Brazil, the mulberry tree - especially the black variety - grows well everywhere, and may be found growing spontaneously in practically all the country's regions.

While the white mulberry is preferred for raising silk-worms, who eat the leaves, the black mulberry is preferred for human consumption for the stronger flavor of its fruit, which are also larger. Also the black mulberry is an ornamental tree which, while not growing to a great height, does have a dense, leafy and therefore shady crown.

All the mulberries are rich in vitamin C and are characteristic in their typical shape, made up of groups of many tiny fruits, rich in water and sugar. Any of the species, picked when well ripened on the tree, are a delicacy for children, adults, and of course, for bird life.

Mulberries are usually consumed *in natura*, often served with cream, but are also delicious when used for pies, ices, preserves, jams, and other sweets, or even transformed into wines, liqueurs and syrups.

MALABAR ALMOND

Scientific Name: *Terminalia catappa* L.
Botanical Family: Combretaceae

Origin: India and Malaysia.
Plant Features: A tree with a straight trunk and a rough bark, producing typically horizontal branches. Large stiff leaves, green when young, turning to yellow and bright red. Yellowish cream flowers on long stems around the ends of branches.

Fruit: Ellipsoid in shape, containing a very hard seed with an edible almond, colored green even when ripe
Cultivation: Indicated for seaside areas, widely cultivated in tropical and sub-tropical countries. Reproduces from seed; should be cultivated in sandy soil.

Beach almond, Indian almond, tropical almond are just some of the names given to this tree, so well-known in Brazil for the cool shade it provides to bathers on the sunny beaches of all the Brazilian coast: a veritable *chapeu-de-sol* (sun shade).

According to Henry Blossfeld, old stories tell that this tree, originally from India, was already common on Brazilian beaches within a century of the first European arrivals. It is supposed that the seeds would have arrived mixed into sands taken aboard in Asia as ballast for Portuguese sailing ships.

As the plant is native to coastal beaches, its seeds can withstand long periods of immersion in sea water, and they remain floating on the waves for days on end until they are thrown up on a beach at high tide. Very easily germinated, they spread rapidly along the endless miles of the Brazilian coastline.

One of the most unusual and unmistakable characteristics of the Malabar Almond is the shape of its crown, which spreads from the top of the trunk, the branches reaching out almost horizontally like the spokes of a wheel.

The large leaves are gathered to form a wide, natural, shady cover. In the winter in the temperate climes the leaves turn yellow, then red, and fall profusely, covering the ground below with a thick organic carpet.

The fruit of the Malabar Almond has an edible pulp which, while rarely made use of, contains a kernel, a beach nut, with a flavor similar to the common almond (*Amygalus communis* L.).

Up to the present, however, the majority of the fruit produced by this beautifully shady tree have served to no more than line the ground with putrid lumps. The only ones to enjoy the fruit from this tree seem to be children, urchins present around all beaches, who, whether seeking food or a pastime, have learned to remove the tasty kernel from its nut.

Currently, in the Federal University of Bahia, based precisely on the observation that the kernel of the sun-hat nut is eaten by the street urchins of Salvador, studies were conducted on the nutritional properties of this fruit, and on the possibilities of using the pulp *in natura* as a food and of using a flour made from the kernel.

BANANA

Scientific Name: *Musa X paradisiaca* L.
Botanical Family: Musaceae

Origin: Asia.
Plant Features: A plant with a juicy, underground stem, the "false" trunk being formed of the overlapping bases of leaves. Large, pale green shiny leaves. Flowers in clusters appearing in pairs from the so-called "heart" of the banana tree.
Fruit: Elongated, soft skin and meaty edible pulp, yellowish in color, depending on the variety.
Cultivation: Propagates from its rhizome, as it has no seeds. May be planted anywhere in Brazil during the rainy season, and produces throughout the year. The most cultivated varieties are: prata, nanica, plantain and pacová. Grows in areas with abundant sunlight and will not tolerate wet soils. One bunch provides from 5 to 40 kilograms, depending on the variety.

The banana probably originated in the hot, damp climate of southeastern Asia, where are to be found the oldest references to its cultivation and the oldest legends built around it. For many, the antiquity and Asiatic origin of the banana are incontestable facts.

It is believed that, throughout its long existence, the banana gradually lost its capacity to multiply from seeds. According to Paulo Cavalcante, this fact is still another indication that man learned to cultivate the banana a long time ago, *"ever since the primordial times of humanity's origins"*.

Today, with the exception of some wild species, the banana can only multiply by vegetative processes, that is, by means of sprouts growing from other plants or scions. If the propagation process is not controlled and there is space, the banana tree may give the impression that it is moving from one place to another, as its sprouts gradually move farther away from the parent plant.

Thus, traveling slowly through time immemorial, the banana has spread over all the tropical and subtropical regions of the globe, where it has become the best-known and most widely cultivated fruit.

Before the Europeans arrived in the Americas, it would appear that some species of native banana or plantain plants were already in existence. Their fruit was not eaten raw, having to be cooked first, and therefore it did not constitute the main diet of the native populations. So it is presumed that it was not until the 15th century that the banana, its cultivation, and its uses were introduced into the American continent.

Today, bananas may be found everywhere in Brazil, especially in the Northeastern and Southeastern regions, where the fruit is mainly grown.

In fact, the banana is the fruit of a plant which could be described as a *"giant herb"*, as poet Paulo Cavalcante says. Actually, this is one of the main characteristics of all the *Musaceae.*

The flowers of the banana are exotic, small and surrounded by a purple bract, when young, known as the *"heart of the plant"*. The fruit, which can be picked when still completely green, grows in great clusters, with a characteristic aspect and shape, one single and abundant time. And the banana fruits are appreciated everywhere in the world.

When unripe the bananas are usually colored green. Their taste is astringent and unpleasant - it is said that a green banana will "pucker your mouth". This is because, before they are ripe, bananas are composed basically of starch and water. So much so that the majority of green bananas are suitable for making an extremely nutritious flour with countless applications as a food, from the preparation of gruel to biscuits. During the ripening process most of the starch in the banana is transformed into sugar, glucose and sucrose. Which is why, in general, the banana is one of the sweetest of all the fruits.

There are many types of bananas. Those which are edible are grouped into varieties according to their consistency and the color of the skin and the pulp, but for each purpose or use one is always better than the others, according to regional and personal preferences.

Table bananas are varieties such as *maçã* (apple), *ouro* (gold), *prata* (silver) and *nanica* (small) - which is actually a large banana, but is so named because

"The most popular fruit in Brazil is the banana. Raw, roasted, stewed; with sugar, cinnamon, in syrup or made into a paste; roasted with cheese; banana meal for children; cream, pies. With honey. As a liqueur. Cooked as an accompaniment to beans, meat, pork... It is an intimate and common fruit, faithful to the poor, savored by all ages and palates. With no work and no complications."

HISTORY OF FOOD IN BRAZIL
Luis da Camara Cascudo

of the low stature of the plant that bears it.

Bananas for frying are varieties such as *banana da terra* and *figo*; nanicas may be cooked but should be breaded to keep them together while frying. Banana-chips, a delicious novelty produced in the north of Brazil, is made from the *pacova* (plantain) variety. Bananas for roasting are also nanica, or in the north, pacovas. The best bananas for cooking with are the plantain types.

The prata variety is often sun-dried, while the figo and nanica may be used for preserves. Banana pastes are made preferably from pratas, but nanicas are also used. Bananas may be made into flour, almost always when they are still green. The fruit also is an ingredient in a large number of typical regional Brazilian dishes. In Rio de Janeiro and Pernambuco, there is a famous stew with many components - meats, roots, legumes and greens - including da terra and nanica bananas. A specialty of southern Minas Gerais is a dish made of nanica banana, corn flour and cottage cheese. On the northern coastline the main dish of *caiçara* cooking is called "azul marinho" (navy blue) - fish steaks cooked with green nanica bananas, accompanied by a *"pirão"* or mush made of the fish juices, mashed cooked bananas and manioc flour. In Bahia, while not using the banana itself, there are dishes such as abará, acaçá and moqueca de folha, which are cooked in banana leaves.

Also worthy of mention is banana brandy, a distillate with a delicate, exclusive flavor, and a specialty of the *caiçara* communities.

The banana is markedly present in home cooking

as well as in popular meals served in bars and restaurants throughout Brazil. Regularly feeding a large part of the Brazilian population, the banana - nanica or prata - is commonly served raw to be mixed into their beans-and-rice or whatever else there is to eat.

Considered, by many, to be the perfect fruit, it is a fruit of many virtues: it ripens slowly after being picked, making harvesting, transportation and processing easier; it is easy to eat, neither too hard nor too soft; it is no trouble to peel; it does not drip juices or sap on one's hands; it has a good flavor, neither too sweet nor too acid; it is not nauseous or indigestible; it is highly nutritious, a few fruit being enough to satisfy one's hunger; it is all used, with no stones, pips or seeds; it has no thorns or spines, fibers or maggots; it grows in any type of soil and produces all the year round.

There is no end to the praises of the banana: Camara Cascudo, for example, states that it has a further utility, this time for anthropology. As it is a plant that is propagated preferably through cultivation; the existence or not of bananas in the diet of an Indian group or community would, he claims, be a sure indication of their degree of isolation. As an example he mentions the case of traveler and geographer Karl von den Stein, who when in the Xingú area at the end of the last century was surprised to note that the native populations there had no knowledge of the best and most beautiful of Brazilian fruits. He said, about the culture of those people, "They have no metals, dogs or intoxicating drinks, not even bananas! Here is a true stone age..."

BURITI

Scientific Name: *Mauritia flexuosa* L. f.
Botanical Family: Palmae

Origin: Brazil - Swampy areas with various vegetable formations.
Plant Features: An elegant palm tree with a straight trunk up to 35m in height. Large leaves arranged in a fan. Yellow flowers in long clusters up to 3m in length, flowering from December to April.
Fruit: Ellipsoid, reddish-brown, having a surface covered with shiny scales. Bright yellow pulp. Hard, oval seed with edible kernel. Fruition from December to June.
Cultivation: Occurs naturally, either singly or in groups, preferably in swampy land, hence its name of palmeira-do-brejo (swamp palm).

Buriti palms and the plains of Central Brazil, immortalized in the literary works of Guimarães Rosa, are an inseparable part of the scrub-covered tablelands.

Wherever a river or stream flows, along its banks, within the tropical fields of the scrub lands and in the so-called *"lavrados"* - enclaves of vegetation similar to that of central Brazil within the tropical forest - of the fields of Boa Vista in Roraima, the riverside forests flourish and in them are the buritis. A short way beyond the forest, and bordering it, are well marked river banks of light-colored sand and low scrub.

In the dense, rich grass of the bottom lands, usually surrounded by open fields, the majestic buriti stands out. The palm has an elegant, straight trunk, crowned with enormous, shiny star-shaped fronds, which form a rounded, uniform and beautiful crown, especially when viewed from below against a clear blue sky.

Seen from a distance, these forests, bottom lands and swamp lands, with a loose, damp soil abundant in buritis, are a sure indication that there is a water course close by, rest and food for the local natives. It is the running water that carries and spreads the seeds of the buriti palm.

Of the buriti - *"a green that perfects and decorates beautifully"*, in the words of Riobaldo de Guimarães Rosa - it has been said and often repeated, since the arrival of the first Europeans with their travelers and

I was homesick for a grove of buritis, among the grasses of a green meadow at the end of the tableland.
My homesickness responded to the wind; a yearning for Minas Gerais. You can picture it: the churning of the wind
among all those buriti fronds when a storm is brewing. Can anyone forget this?
The wind is green. Ai, - meanwhile you can pick up the silence and hold it on your lap.
I am from where I was born. From other lands...

GRANDE SERTÃO: VEREDAS
João Guimarães Rosa

naturalists, that it is the handsomest palm in existence.

Also, in the regions where it grows, the buriti is the most important plant of them all to the local inhabitants, who, inheriting the wisdom of the Indians, have learned to depend upon the plant for much of their sustenance.

The bunches of fruit and fronds they need are gathered at the top, cut close with a sharp knife to avoid damaging the palm. After harvesting, the experienced woodsman jumps, using the wide buriti fronds as a parachute, landing gracefully in the water.

The fruit of the buriti is a small brown coconut that, when unripe, has hard scales which gradually darken as it ripens, and it has a golden yellow-colored pulp. To extract the pulp, it is necessary to first soften the scales in warm water or steam them in leaves or plastic bags. The pulp is used to make sweets and other traditional products, such as the raw brown buriti sugar, always packed in delicate wooden boxes, made of none other than the buriti wood itself; buriti flour, produced from the internal part of the palm's trunk; buriti shreds, made from the shredded sun-dried pulp of the fruit; a buriti dish called *paçoca*, made of shredded pulp, cassava flour and buriti sugar. All these foods are able to remain preserved throughout the dry season, when other foods become scarce.

The pulp may also be frozen and kept for more than a year, and used in much the same manner as the fresh pulp. Today it is also used for producing various types of ice creams, custards, jellies, liqueurs and vitamins with exotic flavors and a high concentration of vitamin C; of all of these modern inventions and discoveries, many were developed in the EMBRAPA research centers.

The buriti does not supply food just for man, it is said that, when the buriti is in season, certain animals eat so much fruit so voraciously that they become heavy and easy to catch. This is the case of the wild pig of Roraima, a species of domestic pig living in the forest, who in this season winds up with its fat stained a strong yellow from the buriti.

But the buriti is more than just food for men and animals. From its pulp, for example, the local population extract a blood-red colored oil used for burns, providing relief from pain and with marked healing properties. This same oil is edible, with a high content of vitamin A. It is also said that the terminal leaf bud also provides a very tasty "heart of palm".

Using the mature leaves, fibers and shoots, as Carmo Bernardes describes it, one can make many things: *"a covering to keep out the rain, basket work for squeezing cassava flour, a hamper for storing flour, a set of baskets... all kinds of mats, mantles and hammocks, cords, sieves, fans and panniers for carrying chickens..."*

Finally, Pio Corrêa says that the buriti trunk, when cut, supplies a pleasant sweet liquid for quenching one's thirst. When fermented, this liquid is transformed into a drink known as buriti wine.

For its beauty and for the many benefits it provides to men and to animals - who also know how to enjoy its fruit - the buriti was the palm tree that most enchanted naturalists Spix and Martius when they first encountered it in the interior of Brazil.

"I ask things of the buriti; and what it replies is: my courage.
The buriti wants it all blue, and will not be separated from its water -
it needs a mirror. A Master is not someone always teaching,
but he who suddenly learns."

GRANDE SERTÃO: VEREDAS
João Guimarães Rosa

BABASSU

Scientific Name: *Orbignya speciosa* (Mart.) Barb. Rodr.
Botanical Family: Palmae

Origin: Brazil - Amazon region and the Atlantic forests of Bahia.

Plant Features: An elegant palm that may reach 20 m in height. Characteristic trunk carrying the remains of old leaves, which have drooped from the apex. Arched leaves, up to 8m in length. Flowers of yellowish cream, agglomerated in long bunches. Each palm may produce up to six bunches between January and April.

Fruit: Oval elongated fruit, brown, appearing from August to January in pendulous bunches. The pulp is floury and oily, involving 3 to 4 oily seeds.

Cultivation: Grows spontaneously in forests in the Amazon region, reproduces through seeds. Each palm tree can produce up to 2,000 fruits, annually, but these do not withstand long storage periods. Hot climate preferable for cultivation.

The babassu is another important representative of the Brazilian palms. Alpheu Diniz Gonsalves stated in 1955 that *"it is difficult to decide what constitutes its greatest exhuberance, whether the beauty of its majestic size or its infinite uses to human beings!"* Nothing could be more true!

The babassu is notable, among other palms to be found in Brazil, the grace and beauty of its characteristic structure: it may reach a height of 10 to 20 meters, with its fronds sticking straight out, only slightly pointing downwards, which gives an even greater appearance of height.

Currently, vast groves of babassu are to be found in Brazil to the south of the Amazon basin, where the damp forest is replaced by the typical *cerrado* (scrubland) vegetation. In the states of Maranhão, Piauí and Tocantins, there are large areas of forest predominantly composed of babassus, often forming dense, dark clumps of trees so closely do they group together. It is quite probable that babassus already grew in this region before the arrival of the Europeans, and were of great significance for the local indian populations.

Camara Cascudo tells us that in 1612 the travelling priest Claude d'Abbeville reported: *Coconuts from East India and Brazil are much appreciated in the areas of Pernambuco and Potiú; however they (the indians) have nothing but the fruit of the palm tree ..."* The palm was probably the babassu, the name in the Tupi language being *uauaçu.*

These ancient babassu groves however would have been diluted among biologically highly complex and varied areas, very different from the present - vast, homogenous babassu groves growing continuously. As the researchers Anthony Anderson and Peter May claim, periodical land clearing and successive fires have been the main causes of the great increase in babassu forests, especially in the Northeastern region of Brazil. These practices, related to an itinerant agriculture, are frequently resorted to for eliminating the babassus themselves, but in the end have a completely opposite effect.

The explanation is that, soon after a large fire, it is precisely the *"pindovas"* or palm tree seedlings that are the first to spring up. This is because it has been discovered that the babassu is extremely resistant, immune to predators of seeds and has a high regeneration capacity and speed. By burning down the babassu and its surrounding vegetation, the main vegetable competitors are eliminated, leaving room for the babassu's subsequent development.

The main product extracted from the babassu that has commercial and industrial value are the seeds contained in the fruit. There are 3 to 5 seeds in each fruit, and these are extracted manually by a traditional and subsistence-level home method. Practically the only living to be made for a large part of the hinterland population in areas where the babassu is to be found, in the state of Maranhão alone the extraction of babassu seeds provides work for over 350 thousand families, especially the women and their children, known as *"quebradeiras"* ("breakers").

Despite countless attempts to invent and implement the use of machinery to perform the task, the breaking of the fruit has always been done in the same, laborious fashion. As the skin or shell of the babassu fruit is extremely hard, the traditional procedure is to position the babassu nut over the sharp edge of an axe held between the "breaker's" knees, and then proceed to hit the nut with repeated hard

blows until it finally breaks in half, when the precious seeds can be collected.

In general, from practically all palms - especially the Dendê, Buriti and Babassu - oils or fats can be extracted for food or industrial usage. Thus the main destinations for the babassu seeds are local crushing plants, producing the raw oil. Comprising roughly 65% of the seed's weight, the oil is a raw material for manufacturing soap, glycerine and edible oil, later to be transformed into margarine, and a cake used to make animal feed and edible oil.

But that is not all. Although it is slow to reach maturity and bear fruit, all of the babassu is made use of, as is true of most palm trees, especially in subsistence or poverty-striken regions. The leaves are a raw material for making utilities - Baskets of all sizes, fans, sieves, mats, fences, windows, doors, traps, birdcages, etc. and as the fundamental raw material for roofing houses and shelters. During times of drought these same leaves serve as food for farm animals.

The trunk of the babassu, if it has rotted, serves as mulch, or, if it is in good condition, can be used for rustic carpentry. From the young palms, when cut down, a palm cabbage is extracted and the fermented sap produces a wine which is much enjoyed in the region.

The green seeds, freshly extracted, grated and then squeezed with a small amount of water through a fine cloth, produce a nourishing milk similar to human milk, according to studies conducted by the Maranhão Natural Resources Institute. This milk is used widely in local cooking as a seasoning for game and fish, replacing coconut milk, and as a mixture for thickening cuscuz made from corn, rice and manioc flour or even drunk as a substitute for cow's milk.

When duly prepared, the nutshells provide an excellent charcoal, the sole source of fuel in various parts of the Brazilian Northeast. The population, making the most of what little they have, frequently produce their babassu charcoal during the night. Burning slowly in kilns covered with leaves and earth, the babassu shells produce a dense smoke which repels insects very efficiently.

Other products with an industrial application to be extracted from the babassu shells are: ethanol, methanol, coke, active carbon, fuel gases, acetic acid and tar.

In spite of its many and varied uses, due to its uncontrolled occurence from an economic and agricultural point of view, the babassu continues to be treated as a marginal resource, and remains only as an integral part of traditional and subsistence systems.

BARU

Scientific Name: *Dipteryx alata* Vog.
Botanical Family: Leguminosae - Papilionoideae

Origin: Forests and scrub lands of Central Brazil.
Plant Features: A tree up to 25m in height with a trunk that may reach 70cm in diameter. Dense, rounded crown. Leaves composed of 6 to 12 bright green leaflets. Small white to greenish flowers appear from October to January.

Fruit: Brown colored fruit with edible seed and pulp, ripening from September to October.
Cultivation: A characteristic plant in scrub lands and forests on dry land areas. Fast-growing, may be grown from seeds. One kilogram of fruit contains roughly 30 seeds.

The Baru, a fruit tree native to the Central Plain in the region of the Mid-western scrub lands, is threatened with extinction.

This scrub land area approximately covers the colossal area of almost two million square kilometers. In Brazil we have scrub land areas in the states of São Paulo, Minas Gerais, Mato Grosso, Mato Grosso do Sul, Goiás, Tocantins, Bahia , Piauí and the Federal District.

Up until the 1960's, the scrub-land areas were sparsely populated, the main economic activity being the extensive raising of cattle on large, practically unimproved properties. The situation today has changed somewhat and, although there are laws designed to protect the environment, much of the forests containing native fruit trees, such as the Baru, are being destroyed. The indiscriminate cutting-down of trees to make room for planting grain and for logging is inexorably affecting the leafy Baru tree.

The wood from this tree, which in some locations may reach as high as 25 meters, is extremely resistant to rot and termites and, for this reason, is much in demand for the manufacture of fences, railway sleepers and boards, and is widely used in civil and naval construction. In all its majesty, however, the tree continues to provide excellent shade, as it is always green, even in the dryest periods.

Also, the Baru, which is an arboreal legume, bears a thin-skinned fruit containing a hard edible seed, the main attraction of the tree for man and for animals. When ripe, the fruit falls freely from the tree and is eagerly eaten up by large animal herds, acting as an excellent dietary complement during times of drought. The kernel of the Baru has a taste similar to that of peanuts and leads the local population to attribute aphrodisiac properties to it, it is said that when the Baru is in fruition, the number of pregnant women increases. What is known is that the Baru has a high nutritional value which, with over 26% of proteins, exceeds that of the coconut.

The Baru kernel may be eaten raw or roasted, when it matches the cashew nut in nutritional value, and serves as an ingredient in recipes for local delicacies. The oil extracted from the nut is of excellent quality and is used by the local population as an aromatizer for tobacco and as a remedy for rheumatism. Despite its many qualities, the Baru is not yet being commercialized and is rarely seen in markets or in the cities.

Along with other native fruit-bearing species in the fields and forests of the Brazilian hinterland, the Baru is being studied and submitted to various experiments by technicians in the EMBRAPA research center ever since the 1980's. The objective, after establishing all the details regarding planting, cultivating and harvesting, as well as the chemical composition of the fruit and physical characteristics of the wood, will be the production of selected seedlings with a view to the rational and commercial exploitation of the plant.

For example, it has already been discovered that the Baru, due to its rapid growth and the quality and strength of its wood, could be of interest to and indicated for reforestation companies.

BIRIBÁ

Scientific Name: *Rollinia mucosa* (Jacq.) Baill.
Botanical Family: Annonaceae

Origin: The Antilles and Mexico.
Plant Features: A small tree up to 6m in height with hairy branches when young. Leaves up to 15cm in length. Flowers are green on the outside and pink inside.
Fruit: Spherical, comprised of several hexagonal parts closely joined, giving it a characteristic aspect. A soft, white, mucilaginous pulp surrounds dark brown seeds. Fruition is from February to June.
Cultivation: The plant grows from seeds and prefers regions with a hot climate and little rain.

The biribá is another important representative of the *Anonaceae* family. Along with the ata, pinha or fruta-do-conde, such as the araticum-do-cerrado or marolo, as the graviola and like the pindaíba, the biribá is known throughout Brazil by various different names. Depending upon the region, it may be an ata, condessa-verdadeira, fruta-da-condessa, etc.

The various types of biribás differ in their weights, amount of pulp, the shape of the fruit, and it is not easy to establish a precise identity for each.

In general, it is a fruit growing on small to average-sized trees, reaching a maximum of 8 meters in height, with a yellow skin covered in excrescences, one for each internal segment of the fruit. The pulp is succulent and free of fibers, colored white or cream, with a pleasant sweet taste. The biribá is usually eaten raw, and is one of the most popular or favorite fruit of the Amazon region.

The importance of the biribá in the north of Brazil is undoubtedly comparable to the importance attained by the commercialization of the sweetsop in the northeast: when it is in season the biribá is to be found on sale everywhere, in all the markets. It's attractive yellow color provides decoration to street corners in large cities, in small riverside markets and the fruit markets like Ver-o-peso in Belem and the Manaus Municipal Market.

While it is rarely cultivated commercially, the biribá tree is to be found in most domestic orchards in the cities of the region and in almost all the orchards of the countryside, its fruit being offered for sale in markets and by street hawkers.

According to Paulo Cavalcante, the biribá originated on the Brazil - Peru border area. For a long time it was thought that it came from the Antilles, having been brought to the Amazon, but, as various similar wild species occur in a number of places, it is difficult to determine its origin. The fact is that today the biribá is widely dispersed from the Caribbean to Southern Amazonia.

BACABA-DE-LEQUE

Scientific Name: *Oenocarpus bacaba* Mart.

Botanical Family: Palmae

Origin: Brazil - Amazon region.
Plant Features: A palm tree, may reach 20m in height with trunk characteristically ringed or scarred where old leaves have detached. Large ascending leaves up to 5m in length. Flowers yellowish-white when young, turning red in time, appear in clusters resembling a "cock's tail" from August to December.

Fruit: Globular, deep purple, with a mucilaginous edible pulp. Fruition from January to July.
Cultivation: Frequently grows in swampy or often flooded regions. Can be grown from seed in sandy, shady soil, rich in organic matter. Can be used as an ornamental.

Oenocarpus distichus Mart.

"The name Pindorama, or land of the palm trees, well suits a country where palm trees abound. The aborigine so named Northern and Northeastern Brazil, as it is there that the carnaubas, babaçus, coconuts, piaçabas and licuri smile. Palm trees predominate there in the coastal panorama and lend their grace to the hinterlands, forming vast islands, where slender trunks support a green dome formed of thousands of fans."

DA INOCÊNCIA DAS PALMEIRAS
F.C.Hoehne

Oenocarpus mapora Karsten

According to Gregório Bondar, of the sixteen species of the genus Oenocarpus, eleven occur in Brazil. The various bacaba palms of Brazil form part of this native environment filled with palm trees and fan-shaped leaves, as described by scientist F.C. Hoehne.

Outstanding among all the noble palm trees of this genus native to Brazil is the bacaba-de-leque. A tree with an elegant bearing, characteristic of the Amazon estuary, according to Paulo Cavalcante, the bacaba-de-leque is common in the forests or dry land coppices, throughout the states of Maranhão, Tocantins and Mato Grosso.

It is easily recognized in the forest due to the peculiar arrangement of its fronds, which are kept to one single plane, forming a true green fan.

The fruits of this solitary palm tree are small and round, "purple-violet" on the outside, as Bondar says . Inside, there is a white- or cream-colored pulp which produces a sweetish edible oil, used in cooking, and equivalent to around 25% of the pulp content.

However, the main use the population has found for the bacaba-de-leque fruit, especially in the Amazon region, is as a form of drink. Cream-colored, with a pleasant flavor, "bacaba wine" is produced and consumed in about the same way as "açaí wine". Both products are similar in terms of nutritional value.

By mixing bacaba wine with the other salty foods served in their daily meals, using it as a moistener to make a mush with local flours, or preparing it together with juices and refreshments, the Amazonian population manages to replace açaí wine perfectly. And synchronized, as the harvesting season for both fruits are complementary, that is, when the açaí harvest is over, during the rainy season, that's just when the bacaba season begins, when its gracious, bounteous clusters hang invitingly from the top of the palm.

The **Bacabinha** (Oenocarpus mapora) is another palm tree of the same botanical genus as the bacaba-de-leque, so much so that the uses its fruit, hearts-of-palm and wood are put to are also very similar. Like the other, it occurs in the Amazon region, especially in the Upper Amazon, as far as Peru.

Like few other bacabas, however, the bacabinha has a peculiarity in that it grows in groves, like the açaí, and in each bacabinha grove a maximum of ten palms may be grouped, besides countless offshoots.

BILIMBI

Scientific Name: *Averrhoa bilimbi* L.
Botanical Family: Oxalidaceae

Origin: Asia.
Plant Features: A tree, up to 15m in height, a trunk with a smooth dark trunk. Leaves are formed of many hairy leaflets. Small, light red aromatic flowers are attached to branches and trunk.

Fruit: Long, lightly-grooved, with a smooth green and yellow surface. The pale yellow pulp encloses 2 elliptical white seeds. Fruition is from September to November.
Cultivation: Indicated for tropical regions. Not well-known in Brazil.

The bilimbi is a fruit very close to the carambola. Slightly smaller, and greener in color, the bilimbi differs from the other in that it is more elongated and has a less sharply-defined, star-shaped cross-section. Its firm pulp and abundant juice contain, as with the carambola, a high content of vitamin C and oxalic acid.

Whether green or ripe, in contrast to the carambola, the bilimbi is usually considered to be too acid and bitter to be eaten raw. Whether processed, salted or sweetened, the bilimbi has the same uses as the carambola: when green it is used for making pickles, and when ripe it is used frequently in recipes for jellies and preserves. Oriental cooking also uses the bilimbi frequently in the preparation of various dishes.

According to Pio Corrêa, the bilimbi has never been found in a wild state and therefore its origins are unknown, as for the carambola. It is generally supposed to have originated in Southeast Asia, in the islands in the region of Malaysia, where to this day it is widely grown and marketed. The name bilimbi is a simplification of the name of the fruit in that region - blimbling asem - which in turn is similar to the name for the carambola - bimbling manis - where asem means sour and manis means sweet.

The English name - cucumber tree fruit - may be a reference to its flavor and appearance reminding one of a small cucumber.

In Amazonia, where it has become thoroughly acclimatized, the bilimbi will have been introduced via Cayenne, in the Guianas, from where comes the name Cayenne lemon, by which it is also known.

BREJAÚVA

Scientific Name: *Astrocaryum aculeatissimum* (Schott) Burret
Botanical Family: Palmae

Origin: Brazil - Atlantic forest.
Plant Features: A palm tree, may grow in clumps, with 1 to 5 trunks, reaching up to 10m in height, densely covered with strong black thorns 6 to 8cm in length. Leaves 2 to 3m long, dark green on top and light green underneath. Small flowers in clusters protected by a spathe covered in thorns and rigid bristles.
Fruit: Egg-shaped or pear-shaped, covered in rigid bristles, with a red seed.
Cultivation: Grows spontaneously in the Atlantic forest, can be cultivated by dividing the clumps or grown from seed.

Among the roughly 3 thousand known species of palm trees, less than 100 produce edible fruit, and not all of these are appreciated for their fruit, nor do they attain economic importance. In spite of this, however, the fruits of these palms, or rather their kernels, form an important part of the diets of native populations where they occur. The brejaúva is one of these.

A palm tree of the genus Astrocaryum, the majority of which are native to Brazil, the brejaúva is to be found practically everywhere in Brazil, from the state of Rio Grande do Sul in the far South, to the extreme Northeastern regions.

One of the main features of the brejaúva is the large number of thorns on its trunk, from which comes the *aculeatissimum* of its scientific name. The thorns are arranged in an attractive pattern and give the palm an ornamental as well as an aggressive aspect. As occurs with almost all palm trees - other than exceptions such as the common coconut, whose fruit is prodigious in food value - one of the main useful features of the brejaúva, besides its kernels and its beauty, is the quality and usefulness of the fibers produced by its leaves.

As in the case of the butiá, mucajá and jerivá, the fruits of the brejaúva are small coconuts, which, when eaten raw, constitute a kind of eating game, a pleasant and nutritious pastime. So much so that games using tops produced by craftsmen from brejaúva nuts became popular, which again makes this palm a natural producer of pastimes, as the children would agree, especially those living in Paraíba Valley region in São Paulo State and the coastal areas of the Brazilian South and Southeast, where the brejaúva is native and abundant. To this day it is still possible to see bunches of brejaúva nuts on sale in public markets in towns such as Guaratinguetá, Taubaté, Pindamonhangaba and Cunha, among others.

Equally famous was - and still is - the passion children have for this little nut, immortalized by the famous Brazilian writer of children's stories, Monteiro Lobato in his *"Sitio do Pica-Pau Amarelo"*. The writer himself was born and raised in the Paraiba Valley, where as a child he must surely have enjoyed brejaúvas.

"Surprise, surprise ..."
"What's the surprise Narizinho? What is it?"
Instead of replying, the girl swept away a pile of leaves near the river stones and revealed to the gang a beautiful bunch of brejaúvas.
"Hooray! Hooray!" shouted Pedrinho, who just loved brejaúvas.
"It was Antonio Carapina who sent them as a present last night".
They all sat around the bunch of brejaúvas and began breaking the nuts on a large flat stone.
"Delicious" exclaimed the boy, greedily eating the delicious soft white pulp of the ripened nuts. "Antonio Carapina has the best ideas in the world..."
CAÇADAS DE PEDRINHO
Monteiro Lobato

CAMU-CAMU

Scientific Name: *Myrciaria dubia* (H.B.K.) McVaugh
Botanical Family: Myrtaceae

Origin: Amazon region.
Plant Features: A small shrub, may reach 3 m in height, stem with a smooth bark. Leaves are reddish when young, then turning a shiny, smooth green. The white, aromatic flowers form clusters of 3 to 4 blossoms.
Fruit: Rounded, colored red when young and dark purple when ripe. A watery pulp surrounds greenish seeds. Fruition is from November to March.
Cultivation: A wild species, occurring predominantly on the banks of rivers and lakes, often with the bottom part of the stem submerged.

According to the results of experiments conducted by the INPA (National Amazon Research Institute) the camu-camu has high nutritious values, particularly a concentration of vitamin C in its pulp that is higher than the acerola, or Barbados Cherry. INPA technicians are also conducting experiments to render cultivation of the plant feasible, by making it more productive. For those who are aware of the amount of ascorbic acid - vitamin C - contained in the acerola and the profits obtained with its economic exploitation, these are important statements.

The fruit of a plant that is native to Amazonia, the camu-camu grows on bushes or small trees and is dispersed throughout most of the region. It can invariably be found on the banks of backwaters, rivers or in permanently flooded regions, where the lower part of its trunk may remain immersed. The camu-camu fruits are small, spherical, the size of cherries, with a tougher skin than that of the acerola, reminding one of the jabuticaba. The skin, bursting in one's mouth, releases its juice from the pulp, which surrounds a single seed. The fruit is reddish, turning to a dark purple when ripe.

The fruit is often to be found in such quantities that the color it gives to the Amazonian river banks attracts the attention of visitors to the region. In Roraima, where it is most profuse, there is even a suburb in the city of Boa Vista that borrowed the name of *caçari*, as the fruit is known in that region.

In spite of its abundance, the native Brazilian has still not learned to make full use of the generosity of this plant. At most, the camu-camu is used as a pastime and appetizer by fishermen during the long hours they spend at the edge of the water, close to the laden bushes. The fruit is also used as a bait for fishing for *tambaqui*, one of the best and more common Amazonian fish.

Today Peruvian Amazonia has some lessons to teach us on using this fruit. There, the camu-camu is rarely eaten raw. As it is very acid, in spite of its sweetness, it is mostly used for preparing beverages, ices, sherbets, jellies, sweets or liqueurs, as well as to add color and flavor to various types of pies and desserts made from other fruit. In all cases, the skin should be added to the succulent pulp of the fruit, as that is where most of the nutrients are concentrated, and it also has an attractive red purple coloring.

The camu-camu is a typically wild species, but with great economic potential giving it the same level of importance as other traditional fruits of the Amazon region, such as the açaí and the cupuaçu.

However, that's not the only place where the camu-camu has a future - in the state of São Paulo, in the Ribeira Valley, a region of swamps and a hot damp climate similar to that of the Amazon, the plant is already being cultivated successfully.

Anacardium occidentale L.

CASHEW

Scientific Name: *Anacardium occidentale* L.

Botanical Family: Anacardiaceae

Origin: Brazil - Northern and Northeastern coastal regions.
Plant Features: The tree may reach up to 10m in height, with a rounded crown proportional to its size, which reaches the ground. Trunk usually twisted and well branched. Leaves are pink when young, eventually turning green. Small, pinkish-white perfumed flowers appear from June to November.
Fruit: The small hard dark-colored fruit is carried by a meaty, succulent well developed stalk, colored yellow, orange or red. The stalk provides the materials for making juices, sweets, etc. The real fruit is the well-known cashew nut, which may grow up to 2cm in length. The fruits ripen from September to January.
Cultivation: Ideal conditions for cultivation are found along the Northeastern coastline. The tree prefers dry soil, and should be planted during the rainy season. It prefers tropical and sub-tropical climates. A 4-year-old tree may produce from 100 to 150kg per year.

The cashew is a rustic tree, spontaneous and native to Brazil more precisely to the sandy coastal fields and dunes that go from the Northeast to the Lower Amazon and has today become disseminated throughout the world's tropical regions. Amazonia appears to have been the warm birthplace of several species of *Anacardium*, which later spread around the world.

Indians speaking the Tupi language, the autochthonous inhabitants of the Brazilian Northeast, were well familiar with the cashew, which was one of their most important foods. Its name, in fact, derives from the indian acaiu, which in Tupi means **nut** (*aca*) **that produces itself** (*iu*). There were the well-known "cashew wars", fought for the temporary control of cashew trees between tribes of indians who would come down from the interior at harvest time, and those living on the coast. It is believed that it was in this manner, through the nuts taken by indians roaming the countryside, that the cashew tree spread over extensive regions of the dry, arid Northeastern hinterlands.

With few requirements as regards the soil, the cashew adapted perfectly to the areas where it was introduced. It flowered and bore fruit year after year, forming extensive groves. There has never been a tree with so much importance and influence in the social life and regional economy of a population. The cashew is present in literature, poetry, popular legends, in speech, children's' games, superstitions, costumes, folklore, medicine and furnishings, not to mention, of course, in the diet and in the culinary art of Brazilian cooks, especially in the Northeast.

The cashew is one of the most interesting fruits to be found; what is commonly believed to be the fruit, popularly called the "pear", "apple" or "banana" - the pulpy portion with an elongated or rounded shape, colored yellow, red or somewhere in between, containing the aromatic and astringent juice, sometimes sour, sometimes very sweet - is merely the swollen stalk or peduncle supporting the real fruit, which is the nut.

Shaped like a small kidney, the nut is the main product in the economic complex of the cashew. When ripe, the nut has a very hard skin, filled with a viscous, inflammable and caustic oil, which surrounds the kernel. During the 2nd World War this oil became a strategic product in industry for its insulating and protective properties. Today, there are over 200 industrial patents using the oil as a component.

Anacardium occidentale L.

"The cashew tree is very much appreciated and there are so many around the coast that they are lost to view. ... The shade provided by this tree is cool and fresh and the fruit is very beautiful."
Gabriel Soares de Sousa (1569)

The indians however were well aware that the best way to enjoy the kernel hidden in the nut was to roast it over a fire. In this manner, the skin and the strong oil it contains are burned away, leaving only the delicious and internationally appreciated *cashew nut*, a luxury item often served salted as an apéritif or in the preparation of sweets and confectioneries.

From the cashew nut, which is rich in proteins, calories, lipids, carbohydrates, phosphorus and iron, an edible oil is extracted which may be used as a replacement for olive oil. In Brazil, in addition to the above, its culinary uses are many. When

Anacardium occidentale L.

green, the large, tender nut, better known as a *maturi*, is the essential ingredient in one Northeastern culinary art, the famous Maturi Fry, with dried shrimp, for example.

From the roasted kernel a very special flour is often made which, when mixed with manioc flour and sweetened is sold in small paper cones, a treat for children. The mixture is also often appreciated when added to cashew juice or to water. It is called *tumbança*.

The succulent, refreshing part, the pseudo-fruit of the cashew, which contains enough vitamins A and C to be second only to the champion Barbados Cherry, has countless uses and, although it has little value on the foreign market, is well appreciated in Brazil. It may be consumed as a fresh fruit beneath the trees or purchased in the streets or on beaches, where it is packed very carefully to avoid bruising. It is commonly served as an accompaniment to a good quality sugar cane liqueur, eaten in pieces or whole, between one sip and another.

At its most popular, as a drink, the cashew is the basis of a number of juices and beverages, when water or milk may be added to the sweetened juice. *Cajuina* is very popular when well iced, being strained juice that has been bottled and boiled in a water-bath. The juice, after having been filtered and cooked and mixed with alcohol becomes *jeropiga*, a cashew wine that may be strong or not. *Mocororó* is the fermented juice, raw or cooked, a finer type of wine.

Northeastern cooking is replete with recipes that call for the fruit, which is a good way to preserve it for the off-seasons. Thus the cashew becomes transformed into all types of sweets, in the form of ices, jams, sun-dried cashew preserves in syrup and crystallized cashew fruit.

When at the time of the great explorations the first Europeans arrived here, they found a promising land of exotic people and fruit, which reminded them of their vision of Paradise on earth, and where the cashew was the forbidden tree. The first marveling descriptions of the tree date back to the 16th century, of the extensive cashew plantations along the coast, the fruit and its uses, written by European travelers.

That was when the cashew began its trip around the world, boarding Portuguese caravels it would land at Mozambique, Angola, Kenya and Madagascar in Africa, and at Goa in India. There the cashews began to grow in profusion in the dry stony land where previously there had been nothing, and soon became completely incorporated into the local life and economy. And it thrived to the point that India today is the world's largest producer and exporter of the highly profitable cashew nuts and oil.

Anacardium humile St. Hill.

Meanwhile, back at its original homeland cashew trees were being replaced, first by sugar cane plantations and a long time later by luxury houses and buildings on the seaside. For many years the possibilities of profitably exploiting the cashew were ignored in Brazil.

The trees that bear the cashew (the meaty peduncle and the nut with its kernel), this tasty, multiple and generous fruit of many fruits in its varying species, may live for over 50 years and present varying features. They range from the majestic Amazonian trees with a 40 meter trunk and small pseudo-fruit (*Anacardium giganteum* Hancock ex Engler), to average size bushes no more than 3 to 4 meters in height (*Anacardium othonianum* Rizz), or 80 cm high shrubs (*Anacardium humile* St.Hill.). In these last cases the nut has the same appearance and uses as the common variety, but the pseudo-fruit is a perfect miniature of the other and slightly more acid. Also known in Brazil as the *cajuí*, *cajuzinho* or *caju-de-arvore-do-cerrado*, it occurs naturally in the scrublands region of Brazil's central plateau.

Anacardium humile St. Hill.

BRAZIL NUT

Scientific Name: *Bertholletia excelsa* H.B.K.
Botanical Family: Lecythidaceae

Origin: Brazil - Amazon region.
Plant Features: A large tree, may reach 50m in height with a base diameter of 2m. Smooth trunk branching only at the upper region, dark, grooved bark, small crown. Leaves are wavy and shiny dark green. Yellowish-white aromatic flowers appear from November to February.
Fruit: Characteristically a spherical capsule with a thick, dark-brown surface. When ripe it releases the seeds by dehiscence of its lower portion. It contains from 14 to 24 seeds, surrounded by a yellow pulp. Ripens from December to March.
Cultivation: Seeds, cuttings or grafts are planted in deep holes with well-fertilized soil during the rainy season. Areas with a hot, damp climate are preferred.

The enormous, fantastic tropical forest of Amazonia has, over the course of history, filled people's imaginations with the existence of fantastic, unknown beings and plants. It is an immense, living and mutating ecosystem of fresh and salt waters, spread over territories with many nationalities and ethnics, entirely covering the Brazilian states of Acre, Amazonas, Pará, Roraima and Rondonia, a large part of the states of Maranhão, Tocantins and Mato Grosso, besides extending into parts of Bolivia, Ecuador, Colombia, Venezuela, Surinam, Guyana and French Guyana.

The Amazon forest, for all its size and biological diversity, is not and could never be uniform. How-

ever, considering the size of the region, one may see distinct forest groupings that are repeated, interspersed and overlapped inside the forest.

First, there are the forests lining the banks of countless rivers, backwaters and lakes, in permanently flooded areas. Second, there are the swampland forests, or *igapós,* periodically flooded during the rainy season, where the waters may rise by as much as ten meters. And, finally, there are the dry land forests, farther away from the flood-prone areas and where, for each hectare, an enormous variety of species of vegetation are to be found, trees of all sizes growing out of shallow, clayey and nutrient-poor ground.

On the forest floor, usually covered with a layer of decomposing leaves and other vegetable matter, the air is hot, stuffy and damp, and there is little light. Among the exposed tree roots countless insects and animals move around in search of food.

Farther up, at 10 to 20 meters in height, in the still fairly gloomy sub-forest, there is a layer of young trees and various palms, their crowns inhabited by parrots seeking refuge from predators.

At 30, 40 or even 50 meters above ground level, capturing more than 90% of the sunlight, there is the upper strata of the forest, crowded with an infinite range of animal life, including insects, birds, amphibious creatures, reptiles and small mammals, all seeking perfumed flowers and succulent fruit. Seen from above, the forest coverage looks like a smooth green carpet hiding exuberant life forms.

It is in this environment that the ancient, noble and majestic nut trees are to be found. The trees are native to the Brazilian Amazon, more precisely to the dry-land forests, and occur spontaneously – together with various species of *sapucaias, maçarandubas, caimitos, abius, sapotas* and *sapotis* – generally in large groups. They are the traditional nut trees.

Also known as the Brazil Nut or Pará Nut, the nut tree is without the shadow of a doubt one of the most important Amazonian trees known, and its exploitation has a fundamental role in the social and economic organization of great forest areas.

The fruits of the nut tree, also called by the natives "*ouriços*", have a very hard woody shell, often weighing as much as two kilos. When these *ouriços* ripen, they drop from the nut tree, and are gathered on the ground. Due to their weight and the height of the tree, the fruit often hits the ground with so much force and at such a speed that, depending upon the consistency of the soil, they may bury themselves.

With the dimensions and shape of a normal human cranium, the nut tree's fruit comprises a strong capsule that does not open spontaneously, housing in its interior a number of seeds, ranging from 10 to 25. These seeds in turn, with sizes varying from 4 to 7 centimeters in length, also have a very hard, rough shell enclosing the sought-after kernel.

The Brazil Nut kernel is very rich in fats and proteins - a veritable "vegetable meat" - the protein contained in only two kernels is equivalent to that in a hen's egg. While it may be eaten raw or roasted, and is a highly appreciated delicacy all over the world, the Brazil Nut also participates as an ingredient in recipes for countless sweets and savory dishes.

By semi-industrial maceration of the Brazil Nut, a fine, edible and very high quality oil is obtained, widely used in the manufacture of pharmaceutical products, cosmetics, fine toilet soaps, etc...

The result of the exploitation of native nut trees, besides providing a living for a large proportion of the population in the Amazon region, has also reached a significant importance in the country's total exports. Recently, however, thousands of nut trees have been cut down by the chain saws of loggers, as the tree provides valuable lumber as well as valuable fruit. Although it is illegal, this practice has caused a significant reduction in native trees, especially in the region known as the Brazil Nut Polygon in the state of Pará.

In contrast, in other locations, such as the states of Acre and Rondonia, very successful experiments are being carried out for the rational exploitation of forest fruits. There, so-called extraction reserves have been established, where workers, organized into co-operatives, gather the nuts and pre-process them, storing large quantities. They are then able to negotiate direct with interested national and foreign groups and obtain the best value for their product.

That is not all. Experiments conducted by EMBRAPA, (Agro-forestry Research Center of Western Amazonia) in Belém, state of Pará, have indicated that, along with other forest essences, the Brazil nut tree is an excellent alternative for the reforestation of areas degraded by grazing or by annual crops. Thus it is hoped that, encouraged by the profits to be gained as much from the production of fruit as by the rational extraction of reforestation lumber, many proprietors and investors will turn to cultivating the Brazil nut tree, leaving the ancient native trees in peace.

CHESTNUT

Scientific Name: *Castanea vesca* Gaertn.
Botanical Family: Fagaceae

Origin: Europe.
Plant Features: A tree, may reach 30m in height, the trunk having a slightly fissured, dark-brown bark. Large, rigid, shiny leaves with serrated edges. Small white flowers grouped around a long stalk.
Fruit: Rounded, with a shell covered in strong sharp spines, containing a large brown nut.
Cultivation: Adapts to temperate climates, preferring calcareous soil. The fast-growing plant is grown in Brazil in the states of São Paulo and Rio Grande do Sul.

The chestnut, known popularly in Brazil as the "Portuguese" or "European" chestnut is well named, as it does indeed come from Europe. Pimentel Gomes states that the chestnut came to Brazil probably from the Iberian or Italian peninsulas, or from both, where it is widely cultivated. A long time ago, lost in history, the tree may have arrived in Europe from Asia.

According to Eurico Teixeira, the chestnut gets its Brazilian name *Castanha* from the town of Castana, located in ancient Thessalia, Greece, where to this day it is cultivated on a commercial scale. In Brazil the plant has adapted well in regions where the climate is temperate to warm with mild summers, such as on the South and Southeastern table-lands.

The chestnut tree is exceptionally large, with a great life-span. According to Pimentel Gomes, it grows rapidly when it is about 10 years old, attaining its maximum height between 60 and 70 years of age, and lives on average for 150 years.

Its fruit contains a delicious edible kernel, covered externally by an aggressive-looking capsule, having long sharp thorns, like a hedgehog. When the fruit ripens, the capsule bursts open revealing the nut, but this does not come away easily, and requires careful manipulation.

These nuts, abundant in Europe, boiled, baked, roasted or reduced to flour, are rich in starch and sugars and for many centuries have formed an important part of the diets of workers and peasants in Europe.

Using the nut whole or chopped into pieces, a series of recipes and techniques were developed in the Old World to use it as a privileged ingredient in the preparation of fine confections. Dried, sugared or crystallized, or even in the form of purées, creams and puddings, the nuts, under their French name of *marron*, are transformed into gourmet's delicacies. It is even said that the flavor of a *marron-glacé* (made from the whole nut, lightly sugared) is the most delicate and best in the world.

CHICHÁ

Scientific Name: *Sterculia chicha* St. Hil. ex Turpin
Botanical Family: Sterculiaceae

Origin: Brazil - Atlantic forest.
Plant Features: A tree, may reach up to 35m in height; its straight trunk may form large prop roots at the base. Leaves characteristically shaped into 3 to 5 lobes, which are shed during the dry season. Small red flowers with an unpleasant smell appear from November to March.
Fruit: Egg-shaped, orange-brown fruit grow at the tips of branches and, when ripe, they open, leaving dark-colored seeds hanging. Fruition from May to September.
Cultivation: Grown from seeds, also used as an ornamental.

The Chichá tree is native to the Atlantic rain forest, occurring naturally along the entire Brazilian southeastern coast line from the state of Bahia as far south as São Paulo. In these regions, in spite of being dispersed over a wide area, the chichá, as a native specimen, is never found in great quantities. Probably its seeds are eaten by wild animals or get buried in the soil waiting for the right moment to germinate.

Reaching a height of some 10 to 20 meters, with a trunk having few branches, all sprouting from the top portion, the chichá is notable in the forest for its height. The ripening of its fruit is a veritable spectacle for nature lovers. High up, at the ends of the branches, from one to four fruits will appear, externally woody and hard.

The brown-colored fruit of the chichá has a rustic, rough, woody aspect and when ripe passes

through an attractive transformation. All the fruit in one same cluster opens up laterally along a longitudinal seam. Each of then takes the shape of a heart, all positioned along the same horizontal plane, showing their velvety black seeds, not more than six for each fruit, adhering to the recently opened mouths. The exposed lips of the mouths are yellow and the inside of the fruit completely hollow. With time, after it has dried, the outside of the chichá fruit turns reddish brown and the whole very pretty and ornamental structure is used as a valuable adornment for dry arrangements.

The seeds are collected after falling to the ground or even, when still on the tree, and cooked or roasted, are edible and while they may be used as food for humans, they are much appreciated by domestic animals who fatten quickly on them.

The chichá produces a light-weight, soft wood, not strong when exposed to the weather, and which, therefore, is more suitable for internal uses, such as ceilings, handicrafts and moldings, as well as for making paper pulp and matches. Due to its fast rate of growth, it is recommended that chichá be planted in reforestation areas for industrial exploitation.

A tall, majestic tree, it has a thick, straight trunk with a light-colored smooth bark. Its exposed tabular roots form what is known as a *sapopemba* or *sapopema*. Rarely cultivated, the chichá is used in some places for the landscaping of city streets and squares, the most famous example of the plant is protected in the center of the Largo do Arouche, in downtown São Paulo City.

CAGAITA

Scientific Name: *Eugenia dysenterica* DC.
Botanical Family: Myrtaceae

Origin: Brazil – Scrub-land areas.
Plant Features: An average-size tree, may reach 3 to 4 m in height, with twisted branches. Shiny green leaves when young are pale green and slightly translucent. Large white aromatic flowers.

Fruit: Round, flattened, pale-yellow fruit containing 1 to 3 white seeds surrounded by a cream-colored, slightly acid tasting pulp.
Cultivation: Occurs wild in scrub land regions.

The cagaita is a native Brazilian fruit, originally from the scrub-land regions.

It has been known for a long time that the *cerrados* or scrublands, which occupy most of the central areas of Brazil, have some of the worst known intertropical soils. A veritable ecological phenomenon, as Prof. Aziz Ab'Saber states, the cerrado vegetation has resisted fires - whether natural or deliberately set by man - being reborn each time from its own ashes.

Recent studies, contrary to popular belief hitherto, have shown that, beneath its appearance of aridity and dryness, the cerrado is surprising in the diversity and wealth of its natural resources. It is also known today that the cerrados, with their distinctive predominant vegetable formations, comprise a veritable natural orchard. Over a hundred native vegetable species in the region provide man, who has learned to use them, with tasty fruits of a high nutritional value. Many other fruits are also appreciated by wild animals.

These species of fruit include the mangaba, a native quince, bacupari, araticum (sweetsop), marolo, burity, piki, cajuí, guabiroba, jatobá, murici, araçá, baru, jaracatiá, curriola, gabiroba and the cagaita, among others. A relative of the *pitanga* (Surinam cherry), the araçá and the uvaia, the cagaita is a small pale yellow rounded fruit. It has a thin skin, an acid flavor and is very juicy, as 90% of the inside is juice.

Despite its pleasant taste and refreshing nature, the people living in the cerrado know that, through a caprice of nature, the cagaita is a fruit to be eaten in moderation. Its scientific name was not given in vain, and, when eaten in excess, the cagaita provokes fermentation, which stimulates the intestinal function and causes a form of drunkenness. On the other hand, an infusion made from the leaves and bark of the tree has the opposite effect and is widely used by popular medicine as an anti-diarrheic.

At EMBRAPA's Agroindustrial Technology Center in Rio de Janeiro, new recipes are being developed and tested, using cerrado fruits. There, roasted baru nut and cagaita juice constitute some of the favorite products among visitors and employees. It has also been found that, while the fruit *in natura* causes an unpleasant reaction, the pulp, used as an ingredient for juices, beverages, ices, sweets, jams and liqueurs, retains no more than its characteristically pleasant taste and scent.

CARNAUBA

Scientific Name: *Copernicia prunifera* (Miller) H.E. Moore
Botanical Family: Palmae

Origin: Northern Brazil and the Pantanal region.
Plant Features: A straight, cylindrical trunk, from 10 to 15m in height, forming spiral patterns on the surface due to leaves that have been shed. Fan-shaped leaves up to 1m in length. Yellow flowers in hanging clusters appear from July to October.
Fruit: A cluster with hundreds of egg-shaped to spherical shiny fruit, green when young and purple when ripe. Fruition from November to March.
Cultivation: The tree grows spontaneously in its natural habitat. It prefers sandy, swampy soils, bottom lands and river banks in hot climates.

The carnauba is one of these palm trees mentioned by the scientist, impressed with the country's exuberance. But it is not just one of them; it is special. So special that it has not escaped the attention of writers such as Mário de Andrade, Guimarães Rosa, José de Alencar and Euclides da Cunha, among others, who in their writings were able to point out man's total integration into the environment in which he lives. And the carnauba, a long-living plant, has been a live witness and active participant in this integration.

When, in the 18th century, the naturalist Humboldt became acquainted with the carnauba in Brazil, he was so taken with the numerous and important uses of the plant that he came to call it the "tree of life". In the Brazilian Northeast, entire dwellings are built from materials furnished by the carnauba, just as from the babassu and the buriti.

The whole fruit of the carnauba is basically eaten by domestic animals; from the pulp of the fruit a kind of flour is extracted as well as a milk, which, resembling the milk extracted from the babassu, may substitute coconut milk. A thorough example of how well man can adapt to subsistence conditions, the seed of the carnauba, which is very oily, when roasted is used locally as a substitute for coffee.

The wood taken from the carnauba's trunk is strong, and may be used for making fence posts, for constructing rustic buildings and as logs. In one piece, the carnauba trunk is used as a pole; sawn up, it provides excellent rafters, beams and laths, and may also be used for shaping turned articles, such as walking sticks and domestic objects. The leaves are used for thatching homes and shelters, and the fibers for making rope, sacking, matting, creels, baskets, hammocks and robes.

Majestic, slim as are most of the Brazilian palms, the carnauba is taller than the babassu and economically more profitable than the buriti. This is because, in addition to the fruit, nuts, log, leaves and the fibers of various uses, the carnauba also provides a wax with a very great industrial importance. The leaves of the carnauba are lined with a waxy covering. The presence of wax on the leaves of some plants, as with the carnauba, is possibly due to its adapting to arid regions, as this waxy layer helps prevent loss of water through transpiration and protects the plant from funguses.

The wax, which is the main product obtained from the carnauba is to this day in the majority of carnauba plantations extracted by extremely rudimentary manual processes. Usually the procedure is as follows: after cutting down, the young palm leaves are layed out on the ground exposed to the sun for several days to dry. As the leaves dry and the wax film covering them is transformed into a floury powder, they are taken to a dark, windowless room of simple construction. There, they are torn apart with large wooden forks and "threshing" begins. The leaves are beaten violently until all the wax has fallen off in the form of tiny white scales and may be separated from the shredded straw. After this powder has settled, it is swept up and taken to a stove with a small amount of water in large kerosene cans. This mixture becomes transformed into a greenish paste, which is poured into a rustic wooden press, which squeezes out the liquid wax, which in turn is poured into clay or wooden molds to harden.

In this manner, a so-called "carnauba wax loaf" is produced, for sale to factories and plants as a raw material for manufacturing a large variety of prod-

"Palm trees - said Martius - abound in the Brazils, flourishing in the sands of the coast, and in the endless fields, rearing up fearlessly and proudly in paués and swamps, spreading their beautiful fronds to throw shade over the green of forests and even living among the arid soil of the scrub lands... Large or small, they spring up everywhere, along the beaches upon hillocks, in among rocks, in fertile or sterile soil".

TOXIC AND MEDICINAL PLANTS AND VEGETABLE SUBSTANCES
F. C. Hoehne (1939) quoted by Gregório Bondar

ucts: shoe polish, lubricants, candles, varnishes, acids, soaps, matches, thermal insulators, record blanks, incandescent lamps, carbon paper, lipstick, etc.

In spite of having so many qualities, of providing so many different products and growing so easily in any tropical climate, it is only in the particularly dry environment of the *caatinga* (scrublands) of the Brazilian Northeast that the carnauba will produce wax in sufficient quantities for economic exploitation. Also, the process for manufacturing the wax involves a series of operations that, due to the rudimentary techniques employed in the Brazilian Northeast, generate a product with a lower quality than that required for industrial processes.

Even so, Brazil is a great producer and exporter of carnauba powder and wax, the state of Piauí being the main producer, followed by the states of Ceará and Rio Grande do Norte. The best and most dense carnauba plantations are to be found in this region which, year after year, is the most subjected to drought in Brazil. There the carnauba plantations extend for miles and miles, bordering the roads leading to the interior of the state or following the courses of the intermittent rivers which may flood when the rains come.

The carnauba is usually located in plenty of space with lots of sun. The fronds are colored bluish-green, are wide and arranged around the tip of the palm in an elegant garland. The short, hard remains of the stalks of old fronds hang in spirals around the trunk, forming clear patterns.

Amidst the ground cover that characterizes all the arid Northeastern region - and which also includes some portions of the states of Maranhão, Paraíba, Bahia and Pernambuco - the immense carnauba trees, both native and cultivated, are very impressive. Who does not know the region or the plant could almost imagine they were looking at a mirage.

COCONUT

Scientific Name: *Cocos nucifera* L.
Botanical Family: Palmae

Origin: Asia.
Plant Features: A palm with a single trunk up to 30m in height, curved or erect. The pendulous leaves are up to 3m in length, with yellowish-green rigid leaflets. The numerous small white flowers form clusters up to 1m in length. The tree flowers all the year round, more profusely in the summer.
Fruit: Egg-shaped, almost spherical fruit, colored greenish to yellow, with a smooth skin, almost 25cm in length and 15cm in diameter, slow to ripen when they become nuts. An abundant pulp up to 2cm thick. The central cavity contains the well-known coconut water.
Cultivation: The plant is propagated by means of seed coconuts derived from mother plants. It prefers sandy soils in hot climates. The following varieties are cultivated: giant (known as Bahia coconut), dwarf, green, yellow, red and hybrid (dwarf X giant).

Some say that the coconut came from India, others that it arrived from the Pacific Islands, still others believe it to be African, while some also say that it has existed since pre-Columbian times in Central America.

What is certain is that Cocos nucifera arrived in Brazil, or more precisely in Bahia, in 1553 aboard Portuguese ships from the Cape Verde islands, where it in turn had also been imported by the Portuguese, according to the traveler Gabriel Soares de Souza.

From the Bahia area it spread all along the coast of Brazil, probably by natural dispersion caused by ocean currents.

In the terra brasilis, the coconut did not immediately reveal all of its potential as a food to the Indians living in the region. According to Camara Cascudo, a little more than 50 years after its introduction, father Vicente do Salvador had observed that large quantities of the great palm trees bearing the coconut were under cultivation, but also stated that the inhabitants of the land merely made use of the nutritious and refreshing water and thin pulp of the green fruit.

According to the author, it was not until the arrival of African slaves, especially those from Mozambique - where the extraction and exploitation of coconut milk were common practices, inherited from far-away India - that the alchemy was initiated, with the creation of delicious and unique dishes from original Afro-Brazilian culinary art.

What dishes of this cuisine could call for coconut milk or even straight coconut? A great many! Among the main dishes, vatapá, caruru de folha, efó, fried maturi, fish and seafood, all types of moquecas, arraia, aratu, shrimp, fish, lobster, oyster and soft-shell crab, chicken xinxim, hauçá rice, and many others. Among the sweets, baba-de moça, cocadas, white, toasted or made from green coconut, for cutting or spooning, the cocada-puxa, quindim, creme do homem, beiju molhado, tapioca couscous, all types of cakes, of cassava, corn, green corn, tapioca, macerated cassava and wheat flour, corn, cassava and tapioca puddings, canjica, pamonha, xerém, munguzá, and banana passoca to name just a few of the possible variations.

Although these are all very specific and typical dishes, many of the principles contained in their preparation have been incorporated into other culinary arts of this continental country, being transformed and adapted in accordance with local ingredients and customs, obviously always maintaining the precious coconut flavor.

For Pio Corrêa, "without a doubt, among the numerous palm trees that are useful to man, this is the most important of them all". It is a fact that, as with all the other plants of the Palmae family, nothing of the coconut palm is lost, but in this case the fruit is essentially and fundamentally the most valuable part.

Before it has completely ripened, while still green, the coconut contains a clear sweet-tasting liquid, known as coconut water. Refreshing, nutritious and therapeutic, coconut water has countless applications, the best of which is to delight the palates of those who know how to appreciate it. It is also efficient as a hydrating serum, and may be used as an aid in the treatment of infantile diseases and frail constitutions.

As it ripens, the quantity of water inside the nut is reduced, at the same time as there is an increase in the thickness and consistency of the pulp. This white, meaty and oily pulp provides the coconut milk so appreciated for cooking. The bagasse, together with whole pulp and other materials is also used for extracting coconut fat, for oils, soaps, candles and margarine. Coconut fiber, which surrounds the nut, is widely used for manufacturing floor mats, carpeting, sacking, brushes, hammocks, etc.

The coconut palm and its fruit are to be found in more than 80 countries throughout the world - in Indonesia, in the Pacific, in Africa, in Central and Southern America and in the Caribbean - and has a significant importance in the lives and the economy of various regional populations. It is at home on a beach, close to the sea and to salt, but this is not a requirement for its successful cultivation. For example, projects are today under way for cultivating coconut palms in irrigated areas of the Northeastern semi-arid sertão, with promising results.

In Brazil, the most common coconut palms are two varieties, the giant and the dwarf. The fruit they bear all have identical characteristics and uses. In the first case, as the giant coconut palm is a long-living plant, with a longevity of up to 150 years, it may grow to a height of 35 meters. This renders harvesting the fruit extremely difficult and dangerous, calling for a great deal of skill, practice and courage on the part of the harvester. However, for all its height, elegance and bearing the coconut palm is one of the most beautiful and ornamental palm trees in existence. When it is from 6 to 9 years of age, the coconut palm begins to bear fruit, which stabilizes at 12 years of age, at an average of 70 coconuts per annum. This is the most common variety throughout the Brazilian northeastern region, which is responsible for roughly 85% of the national production and over 90% of the planted area, mainly in the states of Alagoas, Sergipe and Bahia.

The dwarf coconut palm - introduced into Brazil in 1925 from Malaysia, grows no taller than 10 meters, making it much easier to harvest the fruit. It bears fruit sooner than the giant variety, in its second year after planting, and also has a better productivity, roughly 200 fruits per tree per year. On the other hand it does not live for more than 20 years, far less than the common centenarian coconut palm.

"Leave me in peace with my mourning and my solitude. Don't speak to me of these things, respect my state of widowhood. Let's go to the stove...
Take the grater and two selected coconuts - grate them. Grate briskly, come on, grate; a bit of exercise never hurt anyone (they say that exercise prevents bad thoughts; I don't believe that). Add in the well-grated white mixture and heat it before squeezing, so that the thick, pure coconut milk will run out more easily. Leave the milk to one side.
Once the first, thick milk has been taken out, don't throw the mixture away, don't be wasteful, for the times are not for wasting. Take the same mixture and scald in a liter of boiling water. After squeezing it, so as to obtain the thinned milk.
What's left of the mixture can now be thrown away as it is only bagasse.
Widows are only bagasse..."

DONA FLOR E SEUS DOIS MARIDOS
Jorge Amado

CARAMBOLA

Scientific Name: *Averrhoa carambola* L.
Botanical Family: Oxalidaceae

Origin: Asia.
Plant Features: A tree up to 8m in height, twisted trunk with flexible branches. Leaves form a dense crown. Small white and mauve flowers, frequently attached to branches.
Fruit: Long, oval fruit, with five prominent sections, colored yellow when ripe. Contains two small seeds in each section.
Cultivation: May be cultivated in hot, damp climates. It is grown from the seed and planting should be done during the rainy season.

The carambola is native to tropical Asia. According to Pio Corrêa, although it has never been found in a wild state, its original home was probably in India. Also according to the same author, the plant appears to have been introduced into Northeastern Brazil in 1817, more precisely into Pernambuco. From there, the tree would have spread all along the Brazilian coastline.

Cultivated today in the tropics of both hemispheres - besides Brazil, the main carambola growers are India, Thailand, Israel and some African countries - the carambola is still considered to be a plant for gardens or domestic orchards. Its cultivation on a commercial scale, for consumption *in natura* as well as for the industrial production of extracts, syrups, juices, jams and jellies, is still far less than it could be, which is a waste considering the fruit's qualities and its generosity.

The tree itself, along with its flowers and fruit, has both beauty and charm. When the carambola tree reaches maturity, its branches, which are long and flexible, often touch the ground, almost completely hiding the trunk, inevitably creating a favorable environment for childrens' imaginations, playing games of "hide-and-seek" or "house".

Before being transformed into fruit, the small flowers of the carambola tree, colored violet in the center and whitish around the edges, completely cover the tree in small clusters. Besides delighting the eye of the beholder, these flowers appear to be pleasing to the palate, being used in some countries as an ingredient in salads.

The carambola is an attractive, curious and different fruit. Ever since they are small, looking like waxen miniatures, the fruits maintain their delicate shape: always with five strongly pronounced segments, so that, when cut in a cross-section the slices have the perfect shape of a five-pointed star. Their coloring varies from light green to golden yellow depending upon the fruit's degree of maturity, with an almost translucent appearance.

The carambola's pulp is generally firm, and its fla-

The carambola tree is an ornamental, especially during its flowering and fruition seasons. Its foliage is thick and the leaves have the characteristic of being sensitive, closing one by one at the lightest touch. They also close to sleep every night.

DICTIONARY OF BRAZIL'S USEFUL PLANTS
and cultivated exotics
M. Pio Corrêa (1926)

vor varies considerably from tree to tree and from fruit to fruit, but is usually sweet when ripe and somewhat acid and astringent when still green. In this last state it may be used for making preserves and pickles, having an exotic flavor and a decorative appearance.

The carambola has many ardent appreciators, and the fruit may be eaten *in natura* or in the form of jams and jellies, juices, sweets, ices or as an ingredient for tropical, refreshing cocktails.

The fruit is rich in mineral salts (calcium, phosphorus and iron), and also contains vitamins A, C and a few of the B complex. It is also a natural source of oxalic acid, which in large quantities is harmful to humans, while carambola juice is popularly used to remove or reduce rust stains on clothing and metal objects.M. Pio Corrêa (1926)

CUPUAÇÚ

Scientific Name: *Theobroma grandiflorum* (Willd. ex Spreng.) Schum
Botanical Family: Sterculiaceae

Origin: Brazil - Amazon region.
Plant Features: A tree, 10m in height with flexible branches. Long leaves up to 60cm in length, rust-colored underneath. Large dark red flowers attached to the trunk.
Fruit: A spherical or egg-shaped fruit up to 25cm in length, with a dark brown, smooth, hard skin. Seeds are surrounded by a white, acid and aromatic pulp. Fruition is from January to May.
Cultivation: Can be grown from seeds in regions with a hot, damp climate. It accepts poor soils and planting should occur during the rainy season.

The cupuaçu is an exaggerated fruit: large, heavy, voluminous and over-scented. The leaves of the tree are large, the seeds are many and also large, surrounded by a white pulp, which is acid, having a strong , pleasant aroma. For some, the scent of ripe cupuaçu is somewhat overpowering.

When ripe, the cupuaçu simply falls out of the tree. The fleshy pulp is removed with cutters and the residue around the seed washed. Once processed, cupuaçu pulp is frozen and may be sold elsewhere in the country or overseas. This refreshing pulp is used to make many types of sweets, a veritable confectionery: juices, beverages and ices; liqueurs and flavored liquors; creams, jellies, mousses and puddings; pies, cakes, biscuits and coverings for other sweets; preserves and jams; paste, firm and crystallized sweets; and the famous cupuaçu bonbon.

A relative of the cacao, the cupuaçu also carries the name of *Theobroma*, or food of the gods. If the cupuaçu is externally unlike the cacao in appearance, it is today known that its nuts contain roughly 48% of oil and that it is possible to extract from them a paste somewhat similar to that from which chocolate or cocoa butter are made.

Cupuaçu chocolate may be found processed in several state capitals of the country and is beginning to be exported along with the nature movement seeking natural and Amazonian products, which has effected the number of people interested in its cultivation. For this reason, in some regions, cupuaçu is already being considered to be the true "white chocolate".

In Nova Califórnia for example, on the highway connecting Rondônia to Acre a group of farmers are making a lot of money by cultivating cupuaçu on a commercial basis. Planted together with pupunha which, besides providing the shade the cupuaçu needs, also supplies palm cabbage, oil for soap, flour and feed for animals, the cupuaçu is already supporting countless families, which hitherto had insisted on fighting the tropical climate and, with little success, were planting rice, beans and corn.

The cupuaçu tree is native to the South and Northeast of Eastern Amazonia, especially in the Northeastern region of the state of Maranhão, and is currently growing throughout the Amazonian basin in Brazil and neighboring countries. In all the region, in Amazonian cities or in the interior, it is rare to find a home that does not have one or more cupuaçus in the orchard.

The cupuaçu tree grows spontaneously in dry-land forests, where it reaches heights of over 20 meters. But, easily adaptable, it may be found under cultivation even in Espírito Santo and São Paulo, though not growing so tall.

During recent years it has been forbidden to take cupuaçu seedlings and seeds out of Amazonia, as an attempt to contain contamination in the great cacao plantations from the disease called "witch's broom" which is common in the forest. Even so the cacao is being hard hit by the plague.

PERSIMMON

Scientific Name: *Diospyros kaki* L.
Botanical Family: Ebenaceae

Origin: Asia.
Plant Features: A tree, up to 12m in height with a rounded, well-branched crown. Shiny leaves that fall in winter. Yellowish-white flowers appear in spring and summer.
Fruit: Spherical, slightly flattened in shape, colored orange, pale yellow, dark yellow, pale mauve to dark mauve, depending on the variety. A viscous orange-red pulp. Fruition from February to April.
Cultivation: Adapts to cold or temperate climates, propagates by seed, slips or grafting. Requires deep, damp soils. Strong winds harm fruition.

The persimmon came originally from Asia, or more precisely from China, from whence it was taken to India and Japan. In time, during thousands of years, it spread out over the five continents.

According to Eurico Teixeira, the persimmon has been growing wild in its natural habitat since time immemorial. In Brazil, where it probably arrived at the end of last century, it acclimatized very well and began to bear fruit better than it did in its country of origin, becoming an important product for commercial exploitation. As Pero Vaz de Caminha once remarked "anything planted in this country flourishes!"

Preferring a subtropical climate, the persimmon tree sheds its leaves completely in winter and, although it is not very demanding with regard to cold weather, its production increases considerably during years when winter is colder. The tree withstands hot weather well, provided the winter is cold and arrives on time. For this reason the tree flourishes in the states of São Paulo, Paraná, Rio Grande do Sul and the higher regions of Minas Gerais and Espírito Santo.

Scattered throughout the South, Southeast and some parts of Central Brazil, over a million persimmon trees guarantee a good, high quality harvest for producers, tradesmen and lovers of the fruit. Over half the production comes from great orchards in São Paulo State, especially in regions such as the Paraíba Valley, Campinas, Sorocaba and Greater São Paulo, and are basically destined for the domestic market.

There are many different varieties of persimmon. Pimentel Gomes states that in Japan alone over 800 varieties are catalogued, some descending from very old lines. For Eurico Teixeira "no fruit varies so much as the persimmon in size, shape, color, pulp, flavor, pulp color, number and shape of the seeds, and skin texture and thickness.

Basically there are three large groups or types of persimmon under cultivation in Brazil: those containing **tannin** or ***sibugaki***, almost red in color, require special treatment after picking to become edible, as they leave an unpleasant astringent taste in the mouth, due to the excess tannin they contain. The ***amagaki*** are sweet persimmons with no tannin, a firm pulp, more yellow when ripe, and which require no treatment before being eaten; and the **variables**, which may either have a yellow pulp and no tannin or seeds, or have a dark pulp and possess seeds and tannin.

No matter which variety is considered, the persimmon is composed almost entirely of pulp. With its gelatinous, cold appearance, it concentrates large quantities of carotene (vitamin A) and vitamins of the B and C complex. The pulp of the persimmon is comprised basically of mucilage and pectine, which accounts for the characteristic appearance of the fruit. Its sugar content (14 to 18%), in the highly assimilable form of glucose, exceeds that of most popular fruits.

An elegant, delicate fruit, the persimmon is usually eaten *in natura* with a spoon. Although little known, some recipes exist for making deserts with persimmons, such as cakes, biscuits and mousses.

A popular delicacy among the Japanese descendants living in Brazil, is dehydrated persimmon - which is best made from varieties with a firm pulp, when not too ripe or too unripe - and is practically the only way to preserve the fruit between harvest seasons. This process, like that for any dried fruit has the great advantage of maintaining the nutritious qualities of the fruit without the addition of any harmful chemicals.

CACAO

Scientific Name: *Theobroma cacao* L.
Botanical Family: Sterculiaceae

Origin: Brazil - Amazon region.
Plant Features: A average-sized tree, up to 6m in height, with many branches. Long, pendulous leaves up to 35cm in length. White, yellow and pink flowers in clusters growing from the branches, from December to April.
Fruit: Elongated, with longitudinal grooves, a hard skin colored from pale yellow to dark red, reaches roughly 20cm in length. Contains a mucilaginous white or pink pulp surrounding five rows of red seeds. Fruition from April to September.
Cultivation: Prefers soils that are deep, permeable, fertile or fertilized, requires shade. Planting should be done during the rainy season. Brazil is the world's largest producer of cacao, exported in the form of dry seed.

The cacao tree, not very tall, was born and grew up in shady tropical forests protected from the strong sun. Its fruits, which vary considerably in shape, roughness, appearance and coloring hang from the entire length of the tree trunk, practically entirely within reach of one's hands. Today it is known that the cacao originated in the American continent, probably in the basins of the Amazon and Orinoco Rivers, from where it would have spread throughout the region, and where some plants may still be found in a wild state.

One such species, known as the *cacaurana* or *capuaçu-do-mato*, grows on trees that are very different from the common cacao. Very tall, *cacaurana* trees reach a height of 20 meters and bear a totally reticulated yellow fruit.

Before the arrival of the Europeans in the Americas at the end of the 15th century, however, only those inhabitants of the area between Mexico and Central America made use of cacao in a complete and systematic manner. These were the Maya and Aztec peoples, who domesticated and learned to cultivate the fruit.

Basically, the process used by these ancient populations was as follows: first, the seeds of the ripe cacao fruit were collected and left to dry in the sun for a few days. Once dry, the seeds were well roasted, then ground and re-ground several times in manual stone grinding mills. During this process, water was added a little at a time until a thick paste was formed. This paste was then beaten until it formed a kind of firm cake, which could be stored and shipped. To use the cacao paste as a beverage, it had to be left to soak in water for a few hours - the longer the better.

Thus, **cacao** - pure or mixed with other ingredients, aromatics and spices - was transformed into a drink that became known as **chocolate**, which delighted everyone who tasted it. With the addition of sugar and asiatic spices, the drink found more admirers and conquered the world, but it was not until recently, at the beginning of this century, that Swiss inventor Lindt developed the machinery and techniques required for the industrial production of the creamy, malleable, almost liquid paste that hardens in the molds into which it is poured. For such, he associated to the modernity of the industrial revolution the basic concept discovered thousands of years previously by the tribes of America: the more you grind and beat cacao, the smoother and more palatable it becomes.

From then on, chocolate as a drink may have lost some of its ancestral prestige, but it did become transformed into a matchless delicacy, giving rise to one of the most important and richest industries in the modern world.

By-products previous to the manufacture of chocolate – such as the cocoa butter and oil extracted under pressure from the cocoa seed and the residues of this extraction – are also very important components in the cosmetics and pharmaceutical industries. Also, from the white pulp that surrounds the cacao seeds a delicious, nutritive juice may be prepared, industrialized and frozen with no detriment to its more famous subsequent processing.

Theobroma bicolor Humb & Bompl.

Theobroma cacao L.

For all these reasons, cacao is a fruit with a noble origin and destiny. It's name, for example, stems from an old American legend: *Quatzalcault* was the Aztec god of agriculture who taught his people how to grow and use the cacao plant. Food for the gods, or *Theobroma* in Greek was the name selected by botanist Linneu in 1737, when naming this species of fruit - which also includes the Amazonian cupuaçu - that offers products so agreeable to even the most discriminating tastes.

The fruit of the cacao and the products derived therefrom - especially chocolate - have always, during the last two thousand years in Western history, been intimately connected to economic power and to the satisfaction of sophisticated delights. For the Mayas and the Aztecs it was the object of a cult, of richness and power. These people learned how to domesticate and cultivate the plant for the purpose of extracting the seed, which was given a value as a currency. With the same seed, patiently pounded and transformed into a paste, they invented chocolate, a beverage considered to be aphrodisiac and a food for the privileged few.

A veritable tree with the golden fruit – vegetable gold – cacao signified power and riches to the Spaniards arriving in America, as long as they were able to maintain a monopoly on the production and sale of the fruit. Introduced by representatives of the Church travelling between Europe and the Americas, chocolate was the object of desires, intrigues and mysteries in the bored European courts, especially in France, Spain and Italy as of the early 16th century. These religious missionaries were in many places those most responsible for controlling the cultivation, processing and shipping of the product to Europe.

Shortly after, cacao was also the object of smuggling and pirating by Dutch and English sailors in search of profits. This broke the Spanish monopoly in the product, and enabled the flavors of cocoa and chocolate to spread throughout the then-known world. In Europe the consumption of chocolate as a beverage and chocolate houses became generalized, enormously increasing the demand for the production of cocoa.

That was the situation in the world when the first seeds for widespread cultivation were introduced into Brazil, more precisely in Southern Bahia State, during the 17th century. It was soon discovered that the region's climate and soil were ideal for the establishment of large cacao plantations. The search for gold stopped right there in the shade of the cacao trees.

From then on, until mid-19th century - and for almost 150 years - cacao became transformed into a symbol of status and power, free spending and riches among the "colonels," who established rich plantations in the regions of Ilhéus and Itabuna, in the state of Bahia. During this period, Brazil became the world's largest producer and exporter of cacao, producing enormous wealth and dominating almost alone the entire international market. For a long time, cacao in Southern Bahia remained a profitable activity, requiring little effort other than owning the land, wielding political power and keeping the manual laborers under control.

The last 15 years, however, have seen the total collapse of cacao production in Southern Bahia. The main reasons for this have been a drop in prices on the international market, competition with African countries producing better and cheaper cacao, an end to the possibilities for expanding cultivable land in the region, a lack of investment in modern farming, the depletion and aging of existing cacao plantations, and to top it off, an incurable disease called *"vassoura-de-bruxa"* (witch's broom) which has infested Bahian plantations.

Today, Brazil is still listed as one of the world's main exporters, but cacao is also produced in other regiuons: in Amazonia, in the states of Pará and Rondônia and in the Southeast, for example.

"And I feel the taste, the aroma,
the hot blood of São Paulo in this
small liquid and savory night
that is my cup of coffee."

MARTIM CERERÊ
Cassiano Ricardo

COFFEE

Scientific Name: *Coffea arabica* L.
Botanical Family: Rubiaceae

Origin: Africa.
Plant Features: A 4m-high bush with a straight stem and a rough gray bark. A conical crown with hanging lateral stalks. Leaves are wavy at the edges, colored gray-green when young, then turning shiny green. White aromatic flowers clustered along the stalks attract bees.
Fruit: Egg-shaped, green, then red, then black depending on the stage of maturation. A smooth, shiny skin contains gray, yellowish-white or greenish-yellow seeds surrounded by a sweet white pulp.
Cultivation: The plant prefers areas with a mild climate, and cannot tolerate frost. It requires fertile, well drained and aerated soil, and develops best in shady locations. There are many varieties from which the various coffee flavors and aromas are obtained.

Although there are still some doubts as to the exact location and period in time when the cultivation and systematic use of coffee came into being, it seems certain that Africa was the cradle of the *Coffea* species, while it is also certain that the Arabians were responsible for its discovery.

It seems that the Arabians themselves, during their processes of expansion, were the great disseminators of the species throughout their then-known world. And it is for this reason that one of the better known species, cultivated to this day, the first to be described by the Arabs and to arrive in the European continent, should have been baptized *Coffea arabica*.

It is generally believed that native strains of this species could have been encountered throughout all the equatorial band that encompasses the African continent, from Ethiopia to the Congo, as far as southern Angola.

Starting from its original birthplace, coffee began its great migration around the world. The Arabs cultivated it along the coastal regions of the Red Sea. In 1690, the plant was then taken by the Dutch to the islands of Java, Borneo and Sumatra, in Indonesia, from where it soon traveled to what is now Sri Lanka on the Indian Ocean, then to India and then into other parts of Asia.

In the early 17th century, coffee coming from these regions fetched high prices on the European market and was being widely sold by the Dutch and Venetians, who hastened to add sacks of coffee to the

"Very profound historical studies have enabled us to establish that the the use of a beverage made from the beans of the coffee tree was initiated by the Arabs of Yemen in the seventh century of our era, therefore between the years 600 and 700. "Kahwah" or "cavé" had in those days been brought from far away, from its home in the high, mysterious land of Abyssinia, where its use was unknown".

DICTIONARY OF USEFUL BRAZILIAN PLANTS
and cultivated exotics
M. Pio Corrêa (1926)

"What happened to the forest here? The worker chopped it down.
Where's the worker? Planting coffee.
Where's the coffee? The girl drank it.
And the girl, where is she ? She's in Paris.
Lucky girl.

MARTIM CERERÊ
Cassiano Ricardo

precious spices being brought from the East.

Reputed to be a product with great medicinal properties, a reviver and exciter of the intellect, coffee was introduced into Europe and was increasingly consumed: the coffee beans, roasted and ground, were poured into hot water and drunk in the Arab fashion, which included aromatization with cinnamon and cloves.

Coffee parlors soon sprang up in the larger European cities, where writers, philosophers and artists would gather to savor the drink in its various forms of preparation.

Ever since, coffee has found its true vocation: the aroma and flavor of the dark liquid, sipped slowly, has the capacity of uniting people and, even if only briefly, creating a moment of silence for reflection.

Between 1706 and 1718, a short period considering the distances and transportation difficulties in those days, the Dutch took coffee from Indonesia to America, via European botanical gardens.

Its cultivation was quickly initiated in the French and Dutch possessions in South America and the Antilles, where, for security and monopoly reasons, it was forbidden to sell any coffee that was able to germinate, grow and produce.

Despite all these precautions, less than 10 years later, coffee arrived in Brazil, more precisely in Belém, Pará, in the hands of Sergeant Major Fran-

"One coffee tree in Amsterdam sired the coffee plantations of Surinam, Guyana and Brazil. Again, it was from one coffee plant in Rio de Janeiro that the first plantations in the states of Rio de Janeiro, Minas Gerais and São Paulo originated. A single plant in Jundiaí also originated the coffee plantations of Campinas and neighboring regions. Therefore the Brazilian coffee plantations constituted an enormous family of one single coffee tree, ***C. arabica cv. Arabica****, coffee also known as National, Crioulo and Typical."*

HISTORY OF THE DEVELOPMENT OF COFFEE CULTIVATION IN BRAZIL
Alcides Carvalho

cisco de Melo Palheta, who acquired cuttings and seeds no one knows quite how. In that same year of 1727, coffee began to be cultivated in Brazil, soon spreading to other regions: Maranhão, the São Francisco Valley, Goiás, Minas Gerais, Espirito Santo and Rio de Janeiro.

In those early days however the importance of coffee in the national economy was negligible. In times of meager knowledge and few financial resources, with the oscillations of international prices, it was not easy to establish coffee growing in the country.

A nomadic and demanding crop - *"an octopus with millions of arms"* as author Monteiro Lobato put it in his book *"A onda verde"* (The green wave) - coffee was always on the move seeking new land as it exhausted the soil it was planted in. To this day, this is still one of the main features in the history of coffee plantations - nomadism.

By the end of the 18th century, coffee was arriving in the Paraíba Valley, in the state of São Paulo. This was the beginning of the great coffee saga in Brazil. But it was not until a few years later, in the Campinas region, the gateway to São Paulo's *"red earth"* areas, that the coffee plantations began to show their full economic potential. In a short while, in mid-19th century, São Paulo began to appear on the lists as one of the great coffee producers and exporters.

The plantations began diversifying, governments and producers began to invest in research for improving the species and creating new strains, and in the development of planting, harvesting and processing techniques. Successive increases in the international price of coffee, a consequence of the expanding North American market, caused coffee plantations to extend westward, throughout the interior of the state of São Paulo and surrounding areas.

The volume of capital generated and negotiated with activities in the coffee industry and with the establishment of huge coffee farms, completely modified the way of life in productive regions and led the new *"Coffee Barons"* to prominent positions on the national political scene.

By 1920, Brazil was the leader of the world's coffee exporting countries - a position maintained for many years - which linked its name indelibly with the image of coffee. Since reaching its peak in Brazil, marked by periods of greater or smaller crises, it can be said that coffee-growing had arrived to stay. Continuing with its migratory trend, coffee went through many regions, attaining an impressive peak in some of them, to then decline quickly, as was the case in the state of Paraná, after the great frost of 1975.

Today, coffee is being cultivated in the *cerrado* country of the state of Minas Gerais, where the climate and the soil greatly favor the planting of *Coffea arabica*, which is reaching a very high level of quality, as recognized by the international market.

There are countless varieties of coffee known as Arabica: different types and strains of coffee, with a higher or lower productivity, rusticity and resistance to blights and weather; requiring more or less care; and from which are obtained beans, powders and beverages with highly-varied qualities and prices.

Some of these varieties - often obtained through experiments of selection, crossing, hybridization or by mere genetic chance - were born and raised on Brazilian lands and soils, such as the Botucatu Yellow, Maragogipe Red, Maragogipe Yellow, Caturra Red, Caturra Yellow, Bourbon Yellow, Mundo Novo, Catuaí Red and Catuaí Yellow, to name just a few.

CURIOLA

Scientific Name: *Pouteria torta* (Mart.) Radlk.
Botanical Family: Sapotaceae

Origin: Brazil - Amazon region and Atlantic forest.
Plant Features: A tree, up to 14m in height, a fluted trunk and gray bark. Hard leaves, green on top and rusty-gray underneath. Small red flowers grow from branches from October to November.

Fruit: Egg-shaped, orange-colored, with a whitish pulp enclosing one single purple seed. Fruition from December to January.
Cultivation: Cultivated only in private orchards. Grows naturally on river banks and in swamps. Grown from seeds in clay soils.

Curiola is the designation both of the fruit and the tree, which is well-known over most of Brazil. Native to this country, the tree may be found from Amazonia to Bahia, down to Minas Gerais, Rio de Janeiro, Goiás, São Paulo, and as far south as Paraná.

Of the same family as the abiu and other *Sapotaceae*, the plant grows by preference close to water, along river banks and in bottom lands. Some species prefer the environment of the rain forest in the Amazon regions (*Pouteria torta*); still others prefer the semi-arid "cerrado" country (*Pouteria ramiflora*).

Curiola fruit are usually greenish or orange in color and sometimes even have a whitish coloring, quite rare in nature. The pulp, also white, is edible, with a very pleasant taste. The fruit is usually eaten raw, but technicians of the EMBRAPA, (Farm Research Center of the Cerrado) in Brasília have been testing and approving recipes for jellies and beverages made from the pulp of the fruit.

Besides the areas where the tree has occurred spontaneously, the curiola may also be found in the domestic orchards of those who appreciate the flavor of the fruit and the beauty of the tree. A natural ornamental, the tree is very tall and has a handsome uniform crown, which makes it suitable for shading streets, parks and gardens.

CEREJA-DE-CAMETÁ

Scientific Name: *Malpighia puncifolia* L.
Botanical Family: Malpighiaceae

Origin: Antilles and Amazon region.
Plant Features: A small tree, up to 5m in height with a smooth trunk, a reddish-brown scaly bark and smooth shiny leaves. Flowers form clusters of 6 blossoms, with white, pink or reddish petals.
Fruit: Globular, yellowish-green in color, turning red when ripe. The succulent, translucent pulp usually contains 3 seeds.
Cultivation: Propagates from the seed, adapts best to hot, damp climates and soils rich in organic matter.

According to Paulo Cavalcante, the cereja-de-cametá (literally, Cametá cherry) or "Madagascar plum" (its popular name in Brazil) is cultivated in almost all the tropical region of the American continent.

In Brazil, while it can be found over a large part of the country, the cereja-de-cametá is little known and rarely cultivated. However, in some regions of Tocantins, especially on the banks of that great river, in the city of Cametá, the plant grows abundantly and is worthy of the local population's appreciation.

Originally from the great island of Madagascar, from where it derives its popular name, the cereja-de-cametá grows on a small tree with an attractive crown. The small fruit has a sweetish flavor and a juicy pulp; while not abundant, it is very refreshing.

Curiously enough, due to its scarlet to purple coloring, almost black when ripe, the fruit of this small tree does have the external appearance of a European cherry, hence its other popular name, although the two plants are not in any way related.

DENDÊ

Scientific Name: *Elaeis guineensis* L.
Botanical Family: Palmae

Origin: Africa.
Plant Features: A palm tree, up to 15m in height, with an erect, dark trunk, ringed due to scars left from shed leaves. Large leaves up to 1m in length with bases covered in thorns. Creamy-yellow flowers gather in clusters.

Fruit: Egg-shaped, elongated, angular in shape, yellow or orange in color. The seed completely occupies the cavity of the fruit.
Cultivation: Develops in areas with a tropical climate in various types of soils, but preferably those that are well-drained but retain moisture and are rich in organic matter. Propagates from easily-germinated seeds.

In Bahia, it is frequently said that they eat "oil food", meaning that the food has been prepared with palm oil, (or dendê oil, or "fragrant oil"). Palm oil is an ingredient that is both difficult to use and delicious when used wisely. An ancestral, sacred wisdom, born in the villages of Central Africa.

Originating in Central Africa, the dendê oil palm, named *dem-dem* in Angola, has been introduced, climatized and naturalized throughout all the tropical regions of the world. According to Pio Corrêa, the first dendê seeds were brought to America over 300 years ago, *"undoubtedly by the first African slaves"*.

Camara Cascudo states that *"as was customary in Africa, an African dish was rarely to be found where dendê oil was not present, imparting color and its peculiar aroma and flavor"*. Thus, among all the important delicacies and spices the Africans brought to America, such as bananas, ochra, yams, coconuts, anise, sesame and some of the peppers - dendê came too.

From there, born of the homesickness and fidelity the slaves held for their native lands, in the heat of mulatto and half-cast cooking fires, in Bahia, a new and ingenious adaptation sprung up: to the Portuguese recipes of eggs and sweet Portuguese olive oil was added the African dendê oil, coconut brought from the Indies, indigenous cassava, and the native fruits of the country, the simplicity and sophistication of a cuisine of unique flavors and odors, fundamentally the fruit of miscegenation.

These half-cast flavors gradually conquered the palates of the white masters, and, while not unanimously, the cuisine certainly became transformed and incorporated into the country's culinary art.

Long after, as Paloma Jorge Amado Costa says, dendê dishes began to *"color with gold the pages of books"*, extracting those flavors from the exclusive and selfish dominion of the streets of Bahia to enrich the dreams and imaginations of the world.

Palm oil, the essential raw material for this culinary art form is extracted from a palm tree, the African oil palm. Its fruit, not very large coconuts, are black to begin with, then as they ripen they acquire a coloring that varies from bright yellow to rose-red, passing through shades of orange and rust. The fruits grow in bunches so abundantly that they press together and become deformed, and, for this reason, they do not have a well defined, uniform shape.

The African oil palm is the most productive oil-pro-

ducing tree in existence, providing 10 times more oil than the soybean, 4 times more than peanuts and twice as much as the common coconut. Its fruit provides two different types of oil, each with different applications.

Palm or dendê oil is the product extracted from the pulp of the oil palm fruit. Sweet, with a strong odor and taste and a somewhat dense consistency, in addition to its wide culinary applications, it is also used in the chemical industry and, after refining, deodorizing and bleaching, it also becomes a raw material for the manufacture of mayonnaise and margarine.

The more noble and valuable palm-kernel oil in turn is extracted from the kernels of the dendê fruit and is extensively used in the cosmetics industry and in the production of chocolate, where it can replace cocoa butter, attaining high prices on the international market.

Today in Brazil the major dendê plantations are to be found in the states of Pará, Amazonas, Amapá and Bahia.

If vatapá, heavily spiced with ginger, pepper, peanuts, does not act upon people adding color to their dreams, wanton condiments?
What know I of such requirements?
I am a widow and should not even talk about such things.
A widow over the stove cooking vatapá, weighing ginger, peanuts, peppers, and that's all.
Next take coconut milk, thick and pure, and finally dendê oil, two large cupfuls: the flower of the dendê, the color of old gold, the color of vatapá. Leave them to cook for a long time on a low heat...
On the same low heat my dreams consume me, it is not my fault, I'm only a widow divided into two ..."

DONA FLOR E SEUS DOIS MARIDOS
Jorge Amado

FRUTA-DE-TATU

Scientific Name: *Chrysophyllum soboliferum* Rizzini
Botanical Family: Sapotaceae

Origin: Brazilian scrub-land regions.
Plant Features: A creeper up to 20cm in height. Large, shiny green leaves.
Fruit: Globular, with a yellow skin and white pulp surrounding 1 to 2 seeds. Fruition is from November to January.
Cultivation: A wild species. No information regarding cultivation available.

The fruta-de-tatu is native to the central areas of Brazil, in the savanna regions. A bushy plant, the fruta-de-tatu grows where sparse formations predominate, basically formed of low trees with irregular branches, twisted trunks and thick bark, which do not hide the surface of the earth.

A low plant, rarely reaching more than 30 to 50 centimeters in height, the fruta-de-tatu may also be found in areas where the vegetation is essentially grass and underbrush.

Many of these plants, which appear to be creepers on the surface grow like veritable trees below the ground, in an intricate system of trunks, branches and roots. One of these is the cajuí (*Anacardium humile*), where the apparently independent shrubs to be seen over a wide area above ground are all interconnected underground.

It is in the midst of this landscape that the plant known as *fruta-de-tatu* (armadillo fruit) appears. Abundant in the savanna areas, its fruit, several at a time, grow close to the ground at the base of the shrub.

The fruta-de-tatu is an integral part of this ecosystem and very important in the diet of the fauna in the region, in fact fundamental for their survival. Including the armadillos.

The *tatus* or armadillos are mammals especially well represented by various species, such as the tatu-de-rabo-mole, tatu-canastra, tatu-bola, tatu-peba and tatu-verdadeiro, among others. Although they feed mainly on larvae, worms, carrion, small vertebrates, termites, ants, spiders and other insects, they are also great consumers of ripe fruits that fall from trees during the fruition season. As the fruta-de-tatu grows close to the ground, it saves them the wait for the fruit to fall and aids their search for food.

The fruta-de-tatu has a white, mucilaginous pulp. Its characteristic flavor allows it to be eaten *in natura* or in the form of juices. It is in fact a thin covering surrounding a maximum of two large stones or seeds, which are sucked and then chewed until the pulp covering their surface has been eaten.

Like many other fruits native to Central Brazil, the fruta-de-tatu has some regional importance, as, like the others, it enjoys considerable popular acceptance.

To date, just as with most other fruit, its exploitation has been of an extractive nature, very often predatory also.

FRUTA-DE-LOBO

Scientific Name: *Solanum lycocarpum* St. Hil.
Botanical Family: Solanaceae

Origin: Brazil – Scrub-land regions.
Plant Features: A bush, 3m in height, with twisted thorny branches. Large oval leaves with white hairs, which make them shiny in sunlight. Large blue-to-mauve flowers.
Fruit: A slightly flattened spherical fruit, green when young, turning yellow and very aromatic. The yellow pulp encloses numerous seeds.
Cultivation: Grows spontaneously in scrub lands and by roadsides.

Fruta de lobo (fruit for wolves) is a species of fruit typical of the Brazilian scrub-land regions, although it may be found growing spontaneously in the Northeast or even as far South as the state of São Paulo. The tree is not tall and has irregular branches and a twisted trunk, looking almost like a bush. Its fruit forms part of the diet of birds, rodents and mammals of all sizes. Opinions are divided regarding the utility fruta de lobo has for man. For some, it is a delicious edible fruit with many possibilities. For others, it is no more than its name suggests - fruit for wolves. The local inhabitants in some regions blame it for serious digestive problems and even poisonings.

For F.C.Hoehne, for example, the fruta de lobo has tremendous possibilities for cultivation, as the fruit, in size and shape resembling a large bright green tomato, "is succulent, very aromatic and well suited to making sweets and desserts, and may even be added to peaches for peach jam and to quince for quince paste". Pio Corrêa agrees, stating that the fruit of the fruta de lobo has the "pleasant aroma of apple" and that it contains "abundant sweet pulp". He also said that the fruit may even be served as a table fruit, to be eaten *in natura.* And that is how fruta de lobo continues to be eaten with no harm to anyone's health. Currently, the pulp of well-ripened fruit - as the unripe fruit contains tannin - is used to make delicious jams and jellies with a slightly tart flavor.

The fruta de lobo is so named for a good reason, as it is very much appreciated by the Guará Wolf, the only true Brazilian wolf, close to extinction, inhabiting scrub land regions, and comprises one of the main items in his diet. Furthermore, recent research in the National Center of Genetic Resources and Biotechnology of Planaltina, in the Federal District, indicates that some substances contained in the fruta de lobo are very important for maintaining the biological equilibrium of the species. Also, for the Indians, fruta de lobo has, since pre-Columbian times, been the fruit which occurs in the same habitat as the Guará wolf.

Along with the Guará wolf, however, the tree and its fruit are to be seen less frequently. They are both suffering not just from man's depredations, but also and especially from the great changes that have been happening to their natural habitat over the last decades. The unbridled devastation of scrub lands to make way for farming activities, initiated a long while ago, and more recently the devastation caused by the great plantations of grain and the practice of utilizing native woods for manufacturing charcoal for steel mills, have hit them both hard.

FEIJOA

Scientific Name: *Feijoa sellowiana* Berg
Botanical Family: Myrtaceae

Origin: Brazil - Southern region.
Plant Features: A bush, 4m in height with a trunk well-branched from its base and a dense crown. Olive-green leaves. Flowers with petals that are white on the outside and purple on the inside, appearing from September to November.
Fruit: Egg-shaped, with thick hard skin, colored light green when ripe. A yellow pulp with a flavor similar to that of the araçá, surrounding 20 to 30 seeds. Fruition is from January to March.
Cultivation: Requires clayey soil and rocky terrain in a temperate climate. A spontaneous plant, can be propagated by seed. Can be used as an ornamental.

The Feijoa or *goiaba-serrana* (mountain guava) as it was traditionally known in Brazil in the past, is a close relative to the common guava. It has large fruit as compared to the first and the tree is small, almost frail, according to Pimentel Gomes.

Pio Corrêa tells us that the "mountain guava" is a species that is doubly useful to man. First, the author claims with reference to the plant as a whole, "by the elegance of its appearance, its persistently attractive foliage and beautiful flowers". In fact, whoever sees the feijoa in full bloom in the spring cannot but admire the flowers with their white petals, colored mauve in the center, contrasting with the purple stamens. These fleshy petals, besides being beautiful, are also edible. According to Pio Corrêa, the petals of the feijoa flower - a children's delicacy - have a pleasant sweet juice, delightful to chew on. The feijoa leaves are also most ornamental. Small and straight, they are dark green on top and silvery underneath.

The second great utility of this plant, according to Pio Corrêa, lies in the "excellence of its fruit, with its thick juicy pulp, very aromatic, abundant and tasty". Its flavor, close to the common guava, is considered by some to resemble a mixture of the taste of banana, strawberry, and especially, pineapple, which explains one of its popular names of "pineapple-guava". The same author also points out that the Brazilians of his time did not give the feijoa fruit the appreciation it merited. In fact, in spite of being a fruit with a sub-tropical origin, native to the South of Brazil, in the grasslands extending as far as Uruguay and Paraguay, it was for a long time a forgotten item in Brazilian orchards and palates.

Taken overseas no doubt by some admirer of the savor of its pulp, the feijoa has thrived in foreign soils and climates. Widely cultivated in Europe, since the end of the last century, especially in the coastal region of Southern France, and also since 1900 on the North American West Coast - in California - it has practically disappeared in Brazil.

In 1981, the agricultural company Dierberger de Limeira, in the interior of São Paulo State, brought the fruit back to the Brazilian scene. Taking the initiative of importing from New Zealand three different varieties of feijoa - one of them developed in the United States - they began to produce seedlings for sale and to re-initiated propagation of the species.

"It seems to be certain that this species is native to the ancient region of Caria in Southwest Asia Minor... However, to make it easier to understand, let us simply state that it originated in the Mediterranean basin, from where it was soon taken to other far distant countries. Its use as a human food is so ancient as to be lost in time – it is recorded among the late stone – age lake dwellers and is entwined into Greek mythology."

DICTIONARY OF USEFUL BRAZILIAN PLANTS
and cultivated exotics
M. Pio Corrêa (1926)

FIG

Scientific Name: *Ficus carica* L.
Botanical Family: Moraceae

Origin: Asia.
Plant Features: A vigorously-growing tree, may reach up to 8m in height. A twisted trunk with a smooth gray bark and fragile branches. Leaves divided into 5 to 7 lobes. Very small flowers develop inside the so-called fruit of the fig, while it is still green.
Fruit: The meaty, succulent, pear-shaped edible structure colored yellowish white to purple, known as the "fig" encloses the numerous fruits of the species, which are frequently mistaken for seeds.
Cultivation: Adapts to any type of soil, preferably deep and permeable. Does however require a temperate climate and cannot withstand frost. Propagation is by cuttings.

It is presumed that the first fig trees, with all their history and mysteries, would have arrived in Brazil in the 16th century. Fig trees of all types, many of them unproductive or bearing inedible fruit, are spread out all over the continent.

According to Pio Corrêa however, up until the beginning of this century, planting of the edible fig - *Ficus carica* - was highly dispersed throughout Brazil and the quantities produced were insignificant. Although excellent varieties of figs - largely originated in Spain, Portugal and North Africa - have acclimatized perfectly to different regions of the country, their cultivation did not surpass the limits of urban orchards or farm gardens.

It was only between 1910 and 1920 that fig planting began in the region of Poços de Caldas, Minas Gerais, the initial production basically destined for local consumption and the domestic production of sweets and preserves.

Under the guidance of the Campinas Agronomy Institute, soon after the drop in coffee production in the early 1930's, and often replacing coffee, there was a great increase in fig production associated with grapes in the state of São Paulo, notably in the areas between Campinas, Itatiba, Valinhos and Jundiaí, and between São Paulo and Mogi das Cruzes. Some of these areas still remain very productive to this day.

The fig is often mistakenly considered to be the fruit of the plant. Actually this is no more than a fleshy hollow receptacle or syconium with a thin, soft skin enclosing the tiny true fruits, seeds and remainders of flowers, the whole being entirely edible. The pink or whitish colored flesh is refreshing and melts in the mouth, while the flavor may vary between insipid and very sweet.

As they are very fragile, figs are easily damaged, which makes their packing, storage, transportation and marketing *in natura* difficult. For this reason, since early history, when they were first utilized by man, other ways were developed to make use of their

qualities - *"highly nutritious and innocent"* as Lúcia C. Santos states - and their sophisticated flavor.

Thus, according to their future destination, whether they come from domestic or commercial orchards, the figs have to be picked at different stages of maturity: green, unripe figs are used basically for making preserves; those that are over ripe are used for the production of dried figs; the ripe fruit being used for making fig-paste sweets or for consumption *in natura.*

Depending upon the characteristics of their flowers and their manner of fruition, there are four general types of *Ficus carica*: Caprifigs, Smyrna figs, Common figs and San Pedro White figs, the varieties most cultivated world-wide being the Common type. The same happens in Brazil: the Roxo de Valinhos variety (named after a municipality in the state of São Paulo, where fig production has been abundant for some time) is most cultivated commercially and also belongs to the Common type.

Among the main characteristics of this variety of fig tree is its rusticity which, added to its vigor and good productivity, makes the fig a lucrative and highly attractive crop.

The fig tree flourishes in temperate subtropical regions, but it also has a great capacity for climatic adaptation. Pio Corrêa elegantly exemplifies this quality of the fig tree, when he says that it is able to adapt *"to the most varied and even opposite conditions of existence"* and that, for this reason, it may be found *"from the seaside, in the burning sand dunes of Libya, to the cold highlands of the Andes, at an altitude of over 3 thousand meters"*.

An example of its adaptability in Brazil is the success obtained with crops as far apart as the cold southern climates of Rio Grande do Sul, and the hilly country in the state of Pernambuco, in the hot Brazilian Northeast. This capacity for adaptation is also reflected in the size of the tree, which can vary greatly, depending upon the climate in the region where it was born and the treatment it received. In regions near the Mediterranean Sea from where it came, when fig trees are left to grow freely, they become enormous. On the other hand, when they form part of orchards and are maintained under heavy pruning in the southern climates of Brazil, the fig trees may be trained not to go beyond the size of a shrub.

GUAVA

Scientific Name: *Psidium guajava* L.
Botanical Family: Myrtaceae

Origin: Tropical America.
Plant Features: A tree, 7m in height, trunk having a scaly reddish bark. Leaves are hairy on their upper surfaces when young. White flowers appear from September to November.
Fruit: An egg-shaped fruit, yellowish green when ripe, very aromatic. Abundant pulp surrounds a large number of hard, small, reniform seeds. Fruition is from December to March.
Cultivation: Can be propagated by grafting. A rustic plant, it prefers hot climates but can withstand mild frost. Not demanding as regards soil, but develops best in fertile, well drained deep soils.

Pimentel Gomes tells us that the guava originated in tropical America, especially in Brazil and the Antilles, where it is found in a great number of varieties. Its wide dispersion probably occurred as the result of the irresistible attraction that birds and small animals have for its delicate and penetrating perfume.

In his dictionary, Pio Corrêa states that the main value of this plant lies in its fruit. He claims that it was solely for its qualities of flavor and sweetness that the guava was taken by European navigators from South and Central America to all the tropical countries in the world. It is a rustic plant, growing in any type of soil. Thanks to its extreme vitality and resistance, the guava can also be found and cultivated in subtropical regions except for in very cold climates where there is frost.

Today, the guava is common in Africa, Asia and Oceania, where it grows either sub-spontaneously or under cultivation, occupying extensive areas as in its native Brazil. Eurico Teixeira states that the diversity of Indian names existing for the fruit shows that the guava has had a wide and very ancient distribution: *xalxocotl* in Mexico; *shuinto* in the quechua language of Peru; *guayaba* or *guava* in the Antilles; *araçaíba, araçá-guaçu, araçá-goiaba* or *goiaba* in Brazil.

All these different names given to the fruit in Brazil reminds one of the similarity existing between the guava and the araçá. The two plants are closely related and the flavors of their fruits are very similar. There are over 150 species of *Psidium* distributed throughout America.

In Brazil, the leaves and flower buds of the guava are often used as a home remedy, an infusion is often administered as a cure for intestinal disorders, especially for infants. The thick but tender skin of the fruit has a high tannin content, (from 13 to 18%), which renders it useful in the cosmetics industry in formulations for treating oily skins and for anti-perspirants.

With a white or red colored pulp, containing vitamins A, B and C, the guava is so sweet that, when fully ripe, it can become disagreeably so.

In the old days, a few generations back, the guava was the queen of pastures, back yards and home orchards. The tree would bear so much fruit that there was enough for insects, birds, animals or even for childhood games. And there would still be enough left over for countless sweets, preserves and jams. To this day in Brazil a guava paste or jam of varying consistency, *goiabada* is far and away the most popular Brazilian dessert.

Probably conceived in the great homes of the Northeastern sugar cane plantations, where the guava came into contact with sugar, *goiabada* became a symbol of the country's confectionery art. A combination "so deliciously Brazilian", as Gilberto Freyre says, *goiabada*, cooked with or without the skins of the fruit, and served with cottage cheese or cream cheese are a great delicacy and a consistent food.

Not until the 1970's have guavas been produced in Brazil on an industrial scale. Today, cultivated in great orchards catering to commerce, exports and industries producing sweets and juices, the production of guava has become more sophisticated. The quality and productivity of the trees have increased significantly, allowing the products to comply with the standards and requirements of international consumers.

Through proper management of the plant and pruning at the right times, it has been possible to obtain a much larger, meatier fruit than that produced in a domestic orchard, with the trees bearing more fruit and continuously. The guava is a naturally prolific plant, and when it is part of a well-formed orchard, at an adult age and managed technically, it can come to produce as much as one hundred kilos of fruit per year.

In this manner, the lonely, native fruit has gained the status of nobility, now producing large annual quantities, especially in the well tended orchards of Southeastern Brazil, notably the areas of Ribeirão Preto in São Paulo State, and irrigated plantations in the São Francisco River Valley.

GUABIJU

Scientific Name: *Myrcianthes pungens* (Berg) Legr.
Botanical Family: Myrtaceae

Origin: Brazil - Southeastern and Southern high-altitude forests.
Plant Features: A tree, up to 20m in height with a twisted knotted trunk and a smooth gray bark. Leaves with pointed tips. The cream-colored flowers appear from October to December.
Fruit: Large, spherical reddish-purple fruit with 1 or 2 seeds. Fruition is from January to February.
Cultivation: The plant grows from seeds scattered by birds. Can be used as an ornamental despite its slow growth. 1 kg of seeds may amount to up to 4 thousand units.

Among the many species of the Brazilian *Mirtaceae* family, the guabiju is a rare tree in a continent-sized country. It grows basically in the southern region, from São Paulo to Rio Grande do Sul, where it is native, and can in fact be found in all the area covered by the upper Uruguay River, through northern Uruguay as far as Argentina, Paraguay and Bolivia.

A delicate tree with a stiff appearance, one of the main characteristics of the guabiju is the type of its leaves: small, hard, dark green in color, the leaves terminate in thin, short but sharp thorns like small needles.

The guabiju does not bear fruit until it is 10 years of age, when its fruition is very abundant, though it only occurs every second year.

According to Maria do Carmo C. Sanchotene, the name guabiju comes from the Guarani Indian language, meaning *"fruit to eat"* or *"fruit with a hard rind"*. In fact the fruit, dark with a firm pulp, according to Eurico Teixeira, contains *"a sweet, aromatic, pleasant tasting pulp, albeit astringent"*.

The tasty guabiju fruit is appreciated as much by man as by wild birds and fauna in general, and is a strong attraction for small animals. For this reason, the tree is sometimes to be found planted in domestic orchards.

In spite of all its good qualities, such as its long life, the flavor of its fruit, the delicacy, bearing and beauty of the tree, the guabiju is not cultivated very often, even in areas where it occurs naturally. However, as its roots are not aggressive and it grows very slowly, the guabiju is to be highly recommended for urban landscaping projects, planted along sidewalks and in squares and parks.

GUARANÁ

Scientific Name: *Paullinia cupana* H.B.K.
Botanical Family: Sapindaceae

Origin: Brazil - Amazon region.
Plant Features: A creeper with a grooved stalk, which may reach a great height. Leaves composed of 5 leaflets. White flowers disposed along the stems.
Fruit: The red fruit split open when ripe to reveal black seeds with a thick white aril, resembling small "eyes".
Cultivation: Grown from seeds almost exclusively in Amazonia. A hot, damp climate is required. The soils of lowland areas and sandy-textured soil should be avoided.

Guaraná is the fruit of a vine, which, when growing in the forest, and depending upon light conditions, may associate itself with large trees and inhabit an intricate world of lianas, branches and leaves, climbing to great heights. On the other hand, when growing in an open, sunny place, it remains a creeper. In either case, the characteristic leaves of the guaraná vine are colored a very bright green, and its many small fruits grow in long bunches.

While it is native to the Amazon forest, guaraná has adapted and come to be cultivated in various other regions of Brazil. In spite of this, however, it is still in the forest that the plant can be found in a wild state, especially and in great abundance in the region around the Madeira, Tapajós and Amazonas rivers and the headwaters of the Marau and Andira rivers.

According to Sônia Lorenz, all this region is the traditional territory of the Sateré-Maué Indians, and it was they that first cultivated the guaraná, transforming the wild vine into a cultivated plant, also discovering the techniques required for processing it.

An Indian legend says that the guaraná is a fruit that sprang from the eyes of the son of Onhiá-muáçabê, who knew all the plants and their uses. The legend says that the child had barely learned to speak when it craved for the fruit of an enchanted nut tree that his mother had planted in the Noçoquém - a sacred place, where the rocks talk - and which was under the power of his uncles. It happens that the uncles did not want the boy. One day, while he was enjoying the fruit of the nut tree, the animals of the forest went and told the uncles, who ordered him

"The Andirazes have in their forests a small fruit that they call Guarana, which they dry and then crush, making it into balls, which they prize as much as the white men do gold, and, grated with a stone and drunk in a gourd of water, it gives them so much strength, that, when the Indians go on a hunt from one day to the next, they are not hungry, in addition to which it is a diuretic, reduces fevers, headaches and cramps."

Father João Felipe Betendorf (1669)
quoted by Sônia Lorenz

killed. When the mother arrived it was too late.

So she buried her son's eyes carefully. The desolate tears of the mother made the guaraná spring up out of the earth, from the left eye the false-guaraná was born, which is of no use; from the right eye the true guaraná was born. After a while, from the grave where the boy had been buried, the other animals began to emerge. Eventually a boy emerged, the first Indian of the Maué tribe, who call themselves *"the sons of guaraná"* or, *uaranã*, which in Indian means *"eye of a person"* or *"looking like an eye"*.

In actual fact, the fruit of the guaraná is blood-red in color and, when ripe, the skins burst, revealing a black seed surrounded by a white covering or aryl, looking just like a human eye. The impression is that dozens of little eyes have appeared in each cluster. But the Indians know that when this happens - when the guaraná ripens on the vine - the time for harvesting is past.

The natives' knowledge goes beyond this. To this day, in the region there is a great distinction between the guaraná produced by the Sateré-Maués - considered high quality - and the guaraná produced by the non-Indian population in the vicinity of the town of Maués, of inferior quality. The production processes employed do not always include the knowledge and traditional practices of the Indians, developed and refined over the centuries.

The process is a lengthy one, beginning with the selection of seedlings in the forest, which are then transported to the location where they are to be cultivated. Two or three years later, when the guaraná reaches fruition, the bunches of fruit are harvested *at exactly the right time*, the fruit skins removed and the seeds washed to eliminate the aryl.

After drying, the seeds are slowly roasted for several hours. The roasted seeds are then beaten inside sacks to loosen the skins of the seeds, which are then peeled manually and pounded for a long time, with just the right amount of water, until a paste is formed, which is then molded into "rods" by specialists in the art; these are *"guaraná loaves"*. These loaves are then washed by the women, and smoked for two long months on rough platforms built over a low fire, before they can be considered fit for consumption.

To obtain guaraná powder, the rods are grated, by the Indians with a stone taken from the riverside, and by the native Amazonian with the tongue of the pirarucu fish. The guaraná powder is mixed with water and may be drunk at any hour of the day, or during special ceremonies and cults.

Besides its traditional presentation in rod form, natural semi-industrialized guaraná may be found already in the form of a powder or as a fluid extract or syrup. It is also an important component in formulas for the production of bottled soft drinks.

According to Sônia Lorenz, from literature produced by European travelers and the tales of Amazonian writers, it is known that the production and sale of guaraná has always been intense in the Maués region, attracting the interest of civilized people as well as of the Indians. Also, for a long time a great demand for the product has always been related to its purported medicinal properties.

If popular theories are to be believed, guaraná, when applied to the human body, acts as a stimulant, an intestinal regulator, an antiblennorrhagic, a sudorific, a cardio-vascular tonic, a retarder of fatigue, and even an aphrodisiac. Many of these properties have yet to be tested or confirmed. What is known is that guaraná is a powerful stimulant, containing after processing a high content of caffeine, often more than that in tea or coffee. Caffeine is a drug with countless applications in pharmacy, but when applied without due control, it may produce serious undesirable side effects.

Currently, studies conducted at the Pharmaceutical Sciences College of São Paulo University indicate a possible use of guaraná in cosmetics oriented toward the treatment of oily skins and cellulitis.

GABIROBA

Scientific Name: *Campomanesia xanthocarpa* Berg.
Botanical Family: Myrtaceae

Origin: Brazil.
Plant Features: A tree, up to 15m in height with an erect trunk, lightly-grooved bark and a dense crown. The pale green leaves give off a characteristic aroma when crushed. The small, whitish-cream colored flowers appear from September to November.

Fruit: Round, greenish-yellow in color, with a greenish succulent pulp enclosing numerous seeds. Fruition is from December to May.
Cultivation: Grows wild in forest regions. Propagates by seed, preferring hot, dry climates.

The gabiroba is another plant of the Brazilian *Mirtaceae* family. According to Maria do Carmo C Sanchotene, the word gabiroba in the Guarani Indian language means *"tree with a bitter bark"*. An important element in identifying the species, the gabiroba's bark, in common with most of the *Mirtaceae*, sheds away in strips, leaving large lighter-colored markings over its surface, giving it an attractive aspect.

There are many species and varieties of fruit in Brazil bearing this same Indian name. Some are bushes, others are great trees reaching a height of 8 to 25 meters. It is believed that the gabiroba is native to the mid-western and southeastern savannas of Brazil, where the occurrence of the bushy, wild varieties is especially pronounced and abundant.

The gabiroba is a rustic tree requiring little care and grows naturally even in poor soils. It is widely cultivated in domestic orchards in Southeastern Brazil and may be found frequently in the states of Minas Gerais, São Paulo, Mato Grosso do Sul and even in Rio Grande do Sul, in almost all the forest formations. It also flourishes in Uruguay, Argentina and Paraguay.

When in flower the gabiroba tree becomes completely covered in small white flowers. The fruit is spherical, with a color changing from pale green to yellow as it ripens. Fruition is abundant: sweet and rich in vitamins, the gabirobas are anxiously awaited by both children and birds, and also attract the interest of domestic animals and fruit-eating fish. In addition to being consumed *in natura*, under the tree, the gabiroba may be made into juices, sweets and ices as well as being a raw material for a tasty liqueur.

Due to its shape and size, the gabiroba reminds one of the araçá; however, from tree to tree, the fruit varies considerably. This is because, in spite of the antiquity of its cultivation and the tastiness of its fruit, it is still a species little used in commercial orchards.

It is the fruit of the memories of country children from Brazil's southeast, when seeking gabirobas in the forest was an exciting adventure. As Câmara Cascudo says in his book *"The History of Food in Brazil"*: *"Children have a vocation for finding fruit trees in fruition. They eat more fruit than the adults... Certain fruit seem to be especially for children."* The gabiroba knows all about this!

"There is another tree like an orange tree called a comichã, which every year bears red fruit of the size and appearence of the myrtle berry, which can be eaten whole, spitting out its one black pip, which is the seed of these trees, whose fruit is very tasty."

NOTÍCIA DO BRASIL (News of Brazil)
Gabriel Soares de Sousa (1587)

GRUMIXAMA

Scientific Name: *Eugenia brasiliensis* Lam.
Botanical Family: Myrtaceae

Origin: Brazil - Atlantic rain forest.
Plant Features: A tree, up to 20m in height with a short trunk and a pyramidal crown. Hard, dark-green leaves. Small white flowers appear from September to November.
Fruit: Rounded, flattened at the ends with a smooth dark-purple skin, almost black spotted with red when ripe. The succulent pulp surrounds 2 white seeds. Fruition is from November to January
Cultivation: The plant propagates from seeds and adapts well to any type of climate or soil, withstanding frost well. It grows rapidly.

The name grumixama, according to professor Pirajá da Silva in a commentary on the work of Gabriel Soares de Sousa, is probably derived from Tupi Indian expressions that mean, "*that which grabs when you eat*" or "*fruit which puckers the mouth*", an allusion to the viscous consistency of the fruit.

The grumixama fruit, however, is not just viscous, it is sweet and tasty. Other travelers in other times, such as Antoine de Saint Hilaire, Pizarro and Spix & Martius, were also enamored of this tree and

its flavorsome fruit.

The fruit is small, smooth and shiny, with a very watery pulp and may be found as one of three varieties: dark red, yellow or purple, this last being far sweeter and more tasty than the other two. Also this purple variety, according to confectioner Lúcia C. Santos, may be used to prepare a wine "*as good as it is little-known*".

According to Eurico Teixeira, the red variety, described in 1587, was common only in the northern region of Brazil, especially in the state of Pará. The yellow and purple varieties may be found growing spontaneously over most of Brazil, from southern Bahia down to Santa Catarina, especially in the rain forest bordering the Atlantic coastline.

In fact the *Mirtaceae*, a family to which the grumixama also belongs, are especially important in all the Atlantic Forest, as much for the diversity of species as for the great number of trees to be seen. They are plants usually found in the shade, but they can also adapt well to sunlight. One of its main characteristics is the permanent molting of its bark, almost always thin and delicate, and which, for this reason, displays a pleasing mixture of different tones of brown.

The grumixama is a very ornamental tree. The leaves are large and hard when compared to the size of its flowers and fruit, and they have a shiny upper surface. When the tree grows by itself, its crown is pyramidal and very attractive.

For all these reasons, and due to the delight of its fruit, which attracts man, birds, insects and other animals in profusion, the grumixama is a species that is never absent from the orchards of those who appreciate Brazilian fruit.

GUARIROBA

Scientific Name: *Syagrus oleracea* (Mart.) Becc.
Botanical Family: Palmae

Origin: Brazil - Northeastern and Southeastern regions.
Plant Features: A palm with a straight trunk, may reach 20 m in height, with a curly, deflected crown. Large leaves up to 3m in length. Flowers appear in clusters during spring, until October.
Fruit: Yellowish-green fruit in bunches, with an edible, oily, white kernel. Fruition is from October to February.
Cultivation: The palm is cultivated from seeds, although it grows spontaneously in forests in Southeastern Brazil. It prefers hot climates and well-drained soil.

The guariroba is a native Brazilian palm tree producing the guariroba coconut, and is quite common in an area from Bahia to São Paulo, passing through the states of Minas Gerais, Goiás and Mato Grosso do Sul.

According to Harri Lorenzi, the guariroba plant is characteristic of the drier forests, where the leaves of trees fall gradually, and may be found as much in scrub land as in the savanna, always in large communities and associated with other plant species. But the guariroba, or Gueroba as it is better known, with its elegant stance and thin straight trunk, is a palm tree that stands out, by excellence, in the flat landscape of the fields of Goiás.

In the fields, the presence of a guariroba is a reliable indication of fertile land, and man and animals utilize almost all of the plant. Its green leaves are avidly eaten by farm animals. Its small coconuts ripen and fall to the ground and are an important complement to cattle feed. Also, the local population extract the kernels from the nuts for making home confections. The kernel also contains over 60% of fats, from which an excellent edible oil may be extracted, which is also widely used for soap manufacturing.

However, of all the products extracted from the guariroba, the most important is its terminal leaf bud. Considered by many to be a bitter-tasting vegetable - which in fact it is, when compared to the sweet hearts-of-palm from the Atlantic Forests - the guariroba heart is a delicacy with wide culinary applications, particularly in some parts of Minas Gerais and Goiás.

For example, in the best recipes for a Goiás pie, (a perfect accompaniment to rice with pequi), it is fundamental to include a generous amount of the bitter guariroba heart. A substantial food and a strong seasoner, the filling of this pie, in addition to guariroba, should also include pieces of chicken, preferably the thighs, sausages, potatoes and boiled eggs, either whole or cut in halves, and ripe tomatoes.

Pio Corrêa states that the bitter taste of the guariroba heart disappears easily if it is boiled with a little bicarbonate of soda. But, whoever has tasted it knows how important that faint bitter taste of the guariroba is to the Goiás pie, savored right from the bottom of the filling, mixing with the other ingredients and the special local sauce.

While they used to be well respected by the local population, who were not in the habit of cutting a guariroba tree down just for its deliciously bitter leaf bud, over the last 30 years, with the transformation of the forests into pastures and farm lands, these palm trees have become more scarce, despite policing by IBAMA (Brazilian Institute of the Environment and Renewable Natural Resources), which prohibits the extraction of native palms.

To make the most of the situation and not go without the delicious delicacy, some land owners have begun to cultivate the guariroba for the use of their families. As the plant is easily cultivated and exploited, these plantations rapidly turned into good business ventures. The main, indispensable steps to be taken by anyone wishing to cultivate the guariroba palm for its leaf bud is to obtain a licence from IBAMA, which, among other things, requires that 20% of the natural forest of the property be preserved.

SOURSOP

Scientific Name: *Anona muricata* L.
Botanical Family: Annonaceae

Origin: Central America.
Plant Features: A tree, up to 6m in height with a straight trunk and small crown formed of large leaves. Large pale-yellow flowers.
Fruit: The largest fruit in its family, egg-shaped, a green skin even when ripe, has short, soft, false thorns. A white, sweet, slightly acid pulp surrounds many dark seeds.
Cultivation: A tropical plant, with no special requirements regarding soil, but prefers clayey soil and develops best in the Northeast or North of Brazil, although it will reproduce in subtropical climates. Can be grown from grafts, cuttings or seed.

The graviola, or soursop, may be considered to be the largest, most tropical, most perfumed and most important among all the fruits known generically as *araticuns*, or better, among all the fruits of the *Anonaceae* family. Also, as occurs with most of them, due to the resemblance the various genera, species and varieties have to one another, and due to their wide dispersion throughout all the tropical region, the graviola has various popular names.

Similar to the biribá as far as concerns its external appearance, characterized by its lumpy, scaly points - false spines, short and curved, but soft - the graviola has a less rounded, more elongated shape than the former, and may vary considerably.

The graviola tree is not noted for its abundant fruition, although it makes up for this with the large size of its fruit: exceptions apart - and there have been stories of trees producing fruit weighing almost 10 kilos - a graviola usually weighs, on average, from 1 to 4 Kg.

The graviola should be picked before it has completely ripened, since, due to its weight, it soon falls, breaks up and is spoiled. On the other hand, it should also not be picked too green or it loses its flavor.

Usually the pulp of the fruit is white, juicy, with a perfumed aroma and a pleasant taste; for Pimentel Gomes, its flavor is *"sweet but slightly acid"*. The pulp is also fibrous, not unlike cotton, and contains many proteins, fats, carbohydrates, vitamins C and B, potassium and phosphorus.

Once it has ripened, or if its skin is broken, the graviola decomposes very quickly. For this reason, it is usual for the fruit to be sold in the form of frozen pulp, as it should be processed almost immediately after ripening. Currently, this frozen pulp is available in European and American markets and in the

"Attractive large fruits, they resemble melons in size and are green. And, on the outside, they are marked with scales much like a pine cone. It is a cool fruit and for the hot weather, and, even if a man where to eat a whole graviola, it will do him no harm. The fruit and its pulp inside resemble creams or custards. This, which you eat, melts in the mouth like water, leaving a pleasant taste."

Fernández de Oviedo (15th century)
quoted by Clara Inés Olaya

larger Brazilian cities, and is indicated for the preparation of juices and ices, often mixed with the juices of other fruits, such as lemon, orange, banana, pineapple, etc., and for countless mousses, jellies and deserts.

The graviola is a popular fruit in Brazil and its cultivation is common in domestic orchards in cities and farms in the northern and especially northeastern regions, where the fruit is also being grown commercially. In these regions, the graviola is also eaten, while still green, as a vegetable, and may be boiled, baked or fried in slices.

Northeastern Brazil is undoubtedly one of the locations in the world where the graviola is most sold and consumed. Almost an exaggeration, when it is in season the graviola is to be found in all back yards, on all market stalls, on all tables, in all mouths, for the greater pleasure of all concerned.

Paulo Cavalcante states that it would appear that the graviola tree exists in a wild state in the Amazon region. The herbariums of the Emílio Goeldi Museum, in Belém, Pará, have several native specimens, with a classification very close to the graviola, collected by various researchers in different locations in the region. However the same author and others believe that the graviola comes originally from the islands in the Caribbean, having been found wild on the islands of Cuba, Santo Domingo, Jamaica and other smaller islands. From there the fruit may have migrated to Amazonia.

In the Antilles, the soursop is known as *guanábana*, from the Taino language that was spoken on those islands, and the same language that originated the word *anon* to designate the sweetsop. As writer Clara Inés Olaya tells it, it was also there that chronicler Fernández de Oviedo encountered the fruit for the first time and recorded its existence.

Centuries later, the graviola is now to be found spread around various regions of the world, where in many of them the plant has become sub-spontaneous. In countries such as Hawaii, for example, the development of techniques for planting and processing the pulp with its exotic and delicate aroma and flavor has transformed the graviola and its by-products into sought-after delicacies within reach of everyone.

INGÁ-CIPÓ

Scientific Name: *Inga edulis* Mart.
Botanical Family: Leguminosae - Mimosoideae

Origin: Brazil - Amazonia.
Plant Features: A large tree, may reach 15m in height. Leaves are divided into 6 to 8 leaflets, attached to a hairy, rust-colored stalk. Flower clusters are greenish-white, flowering almost all the year round.
Fruit: Long and linear, reaching as much as 1m in length, brownish-green in color. A white, fibrous pulp surrounds black seeds. Fruits are borne almost all the year round.
Cultivation: Cultivated from seed. The plant prefers sandy soils close to rivers. With a rapid growth, this species is the best known of the *"ingás"*.

Within the dense, rich forest, among the river backwaters, there are a countless number of trees known as ingás. At the water's edge, the ingás join the bacabas and clumps of açaí palms full of fruits and food.

In the dry land forests, full of gigantic trees hung with festoons of lianas, with exposed roots, leaves, fruit and animals seeking them, other varieties of ingá are to be found. All of this surrounded by heady perfumes, magic and silence, broken only by

"I slide through a labyrinth
with heavy trees seated in the dark.
Hungry roots bite into the ground.
Dirty carobas raise their skirts
like clusters dripping mud.
Long-legged cabbage-palms
wave their fronds slowly in the heavy air
like the legs of a spider pinned to a stalk.
Miritis open their great fans slowly."

COBRA NORATO
Raul Bopp

the sudden flapping of birds.

According to Pio Corrêa, the Indian name Ingá signifies "*soaked*" or "*steeped*", possibly due to the consistency of the fruit's aryl, a watery pulp surrounding the seeds, and it is the name of over two hundred species of the *Inga* genera, in the *Leguminoseae* family.

Not all are native to the Amazon forests like the ingá-cipó, but in general ingás prefer to inhabit the forests along the banks of backwaters and Amazonian rivers. When they occur in other regions, they are still characteristic of riverine forests.

As with all the Brazilian ingás, the ingá-cipó fruit is a bean. In this case in a large green pod. The main characteristic of this species - and which makes it stand out from the other ingás - is that its bean pod may reach a length of 1 meter without breaking, and as it is also spiral in shape, this may be the reason for its name *cipó* (literally a liana or vine).

The pod contains shiny black beans or seeds surrounded by a white, slightly fibrous, soft and sweet-tasting aryl. These seeds are sucked and spat out. Although the pulp has nutritious properties, the fruit is eaten by the Amazonian population more for fun than for food.

Ingá beans are frequently found for sale in the markets of Amazonian cities, and are easily transported from forests or other cultivation areas to market without spoiling. Very popular throughout Amazonia, the ingá-cipó is often cultivated around homes and elsewhere and is common in the forests in a sub-spontaneous state. It is also common practice to use ingá-cipó trees to provide shade over coffee plantations in the region.

JARACATIÁ

Scientific Name: *Jaracatia spinosa* (Aubl.) A. DC.
Botanical Family: Caricaceae

Origin: Brazil - widely distributed.
Plant Features: A large tree, may grow up to 20m in height, with trunk and boughs covered in spines and a rough bark. Leaves are composed of 5 to 8 leaflets. Small, greenish flowers appear from September to October.

Fruit: Oval or rounded in shape, colored orange-yellow when ripe. Yellowish or reddish pulp depending on stage of ripening, sweet, surrounding numerous seeds. Fruition is from January to March.
Cultivation: A wild plant native to fertile soils. Propagated by seeds.

The Jaracatiá tree is native to Brazil, where it is widely distributed over a large part of the country. Traveler Gabriel Soares de Souza had already recorded its presence in the country in the mid-16th century, calling it *"indigenous to Bahia"*. According to Harri Lorenzi, the jaracatiá may grow in several different forest formations, from southern Bahia to Rio Grande do Sul, including Minas Gerais, Goiás and Mato Grosso do Sul.

Another of the typical fruits of the Brazilian savanna areas, the jaracatiá occurs naturally in calcareous forests and the so-called riverine forests, growing adjacent to or along the banks of water courses. Or rather it used to grow. Today the jaracatiá is in danger of extinction, only to be found on traditional farms, where native plants and trees have been preserved, or else in the gardens of the Cerrado Farmland Research Center, EMBRAPA of the Federal District.

The jaracatiá is an interesting fruit. Although it is known regionally by such diverse names as a tree-papaya, thorny papaya, wild papaya, and deer's papaya, the sweet flavor of its pulp is far different from that of a common papaya.

To begin with, the jaracatiá grows on very large trees as compared to the papaya. Reaching heights of between 10 and 20 meters, the jaracatiá tree is very ornamental, since the whole plant - with its tapered trunk, well spaced out thorny branches, and a crown bearing few, but very beautiful leaves with an unusual format and organization – composes a bizarre and elegant picture, all looking upwards.

Secondly, the papaya and jaracatiá fruits, although having a similar appearance, format and coloring, have a different type of pulp and seeds. Due to the gelatinous consistency and appearance of its pulp, the jaracatiá fruit is most suitable for making jellies, as opposed to the papaya, which is best for sweets and preserves.

Jaracatiá fruit is avidly sought after by birds and monkeys, and may also be eaten by man *in natura*.

In Pio Corrêa's dictionary, there is one reservation regarding overindulgence of the jaracatiá: the fruit contains a milk which has a purgative and irritating effect, especially when it is green, and may cause general sickness and fever. It is said that this sickness was experienced a number of times by soldiers during the expeditions to the interior of Brazil commanded by Marshal Rondon at the beginning of the century. It would seem that the method most indicated, to neutralize somewhat the effect of jaracatiá milk, is to boil the fruit as a legume. In northeastern Brazil, in order to eat the fruit raw, it is customary to score the skin with a fork and leave it to drain overnight.

Eugenia malaccensis L.

JAMBU

Scientific Name: *Eugenia* sp
Botanical Family: Myrtaceae

Origin: India.
Plant Features: A tree, up to 15m in height, has a dense, conical crown with many branches. Shiny green leaves. Large aromatic flowers varying from white to pink and purple according to the species.
Fruit: Egg-shaped, colored white, green, pink, yellow or dark red, with a succulent white pulp surrounding spherical seeds.
Cultivation: Cannot withstand frost and develops in any type of permeable, deep soil. Cultivated in almost all Brazilian regions with a hot damp climate. Propagation is by seed. The plant can produce for over 20 years.

Eugenia jambos L.

"This fruit I show you is highly esteemed in this country:
it came from Mallaca a short while ago, as there are many in those parts.
But I tell you this apple is the size of a duck's egg or somewhat larger;
its color is red and white and it smells like rose water, so that it is pleasant to both senses"
"I believe", says Câmara Cascudo, referring to the above text
"that this fruit is the red jambo".

HISTORY OF FOOD IN BRAZIL
Camara Cascudo

Red jambo, white jambo, wild jambo, rose jambo and yellow jambo – delights brought to Brazil from India and some Malaysian islands.

Although the beautiful jambo trees are not native to the Americas, they have acclimatized very well to American tropical conditions, and appear to have originated in this continent. In the northern, north-eastern and southeastern regions of Brazil, few orchards are without at least one jambo tree, and no one asks whether it is native or imported.

Although they are all called jambos, they are not really varieties of the same fruit. Their relationship may possibly be the same as that between a jabuticaba and a pitanga, also *Mirtaceae* as are all the jambos.

The better known red jambo (*Eugenia malaccensis*), also known as the ohia or mountain apple, is a large tree with large leaves and a pyramidal crown. The tree is extremely ornamental and offers beauty, good shade and sweet fruit to those fortunate enough to possess one. Its many beautiful flowers, are colored purple, pink and lilac, and when they fall they form a *"purple carpet with a beautiful effect"*, as Pio Corrêa so aptly describes it.

The fruits are red-skinned, with a white pulp having a sweet flavor reminding one of a pear with an added flower-like perfume. The skin is thin, and the not very succulent pulp is firm, requiring to be chewed quite hard when eating it.

The red jambo fruit can be used to make a special and delicious preserve: the fruit is partially peeled and cut into halves. These are then placed quickly in boiling water and next cooked in a sugar syrup. In addition to being a delicious dish, it also has a beautiful pink color, the color of a peach dessert. The recipe comes from Dona Chloé Loureiro.

The rose jambo, also known as the yellow jambo or rose apple (*Eugenia jambos*) grows on a smaller tree than the red variety. The leaves are also smaller and thinner. When in flower, the abundant, large, white but delicate blossoms cover the crown of the tree, to then fall to the ground, providing a pretty spectacle.

The fruits too are smaller, in their external appearance far more resembling an araçá than a red jambo. The color is also lighter, varying between shades of rose and yellow; in the words of Pio Corrêa, their color may be *"rose-yellow or rose-white or purple"*. The floury consistency of its pulp reminds one of a juiceless apple similar to the red jambo, but the flavor is milder and has a strong rose-like perfume - hence its name.

The white jambo (*Eugenia aquea*) is the most interesting and exotic of all the jambos, as the fruit, more similar in appearance and form to the rose jambo than the red, but white in color, is something quite rarely found in nature. When describing it, Pio Corrêa, stated that the fruit *"is very handsome, as if made of porcelain or wax"*. In spite of having a milder and less pronounced flavor than other jambos, the white jambo still has many admirers.

Eugenia aquea N. L. Burm.

JAMBOLAN

Scientific Name: *Eugenia jambolana* Lam.
Botanical Family: Myrtaceae

Origin: India.
Plant Features: A tree, up to 10m in height with a large, many-branched crown. Smooth, shiny leaves. Red flowers with rounded petals characteristically in the form of a hood.

Fruit: Egg-shaped, small, a reddish purple or almost black when ripe. A meaty pulp surrounds one seed.
Cultivation: Cultivated from seeds, it grows well in any type of permeable, deep soil. Prefers hot, damp climates, especially in coastal regions.

The jambolan is a large, branchy and very beautiful tree of the *Mirtaceae* family, which produces some of the most Brazilian of all the fruits: jabuticabas, cambucis, pitangas, cambucás, gabirobas, cambuís, araçás, grumixamas, guabijus and so many others, but this one, according to several authorities on the subject, such as Eurico Teixeira and Pimentel Gomes, comes from India. This same author also states that besides being eaten raw, the jambolan is also used for making confections and sweets, something that, the Brazilian has not yet learned to do.

Jambolan fruit, in spite of being somewhat astringent, is pleasant tasting, but as compared to the exquisite flavor of many other Brazilian *Mirtaceae*, its flavor is not all that remarkable. It is not large, purple-colored, almost black on the outside, with a not very juicy pulp and a single seed, large when compared to the overall size of the fruit.

Some see in jambolan fruit a certain resemblance in size and in color to an olive, and, for this reason, the fruits are frequently referred to as "olives", especially in the north of the country. In fact, in that region the plant has adapted so well that it has become a sub-spontaneous species. In Pernambuco, for example, the leafy "olive" trees may be seen for miles along highways leading to the interior of the state.

While they are avidly sought by children, the jambolan fruits have one drawback: the purple-colored skins and pulp leave indelible stains on hands, clothing, street sidewalk, and automobile paintwork, so that, in spite of its ample fruition, the jambolan is not popular in gardens or public spaces. Pio Corrêa says that, at one time in Belo Horizonte, the capital of the state of Minas Gerais, the jambolan was tried out for urban arborization, but this was hastily abandoned for the above reasons, once the plant began to bear fruit.

For children, however, the drawback becomes fun: beneath a jambolan tree, the game is to eat as much fruit as you can, to see who has the most purple-hued tongue!

"Behind the school there are jabuticaba trees.
We study when we have to, but afterwards off we go to steal
jabuticabas. Jabuticabas have to be eaten up in the tree.
The act of stealing wipes out the theft.
Conscience lighter than a wing.
I descend and go home with empty hands."

MENINO ANTIGO
Carlos Drummond de Andrade

JABUTICABA

Scientific Name: *Myrcia cauliflora* Berg
Botanical Family: Myrtaceae

Origin: Brazil - Atlantic forest.
Plant Features: A tree, up to 8 m in height with a pyramidal crown. Leaves are red when young, turning green. White flowers grow directly from trunk and branches, flowering twice a year, from July to August and from November to December.
Fruit: Globular, deep purple fruit with a sweet, white pulp surrounding 1 to 4 seeds. Fruition is from August to September and from January to February.
Cultivation: The jabuticaba tree prefers deep soils, rich in organic matter, and requires a great deal of water. It will develop in any type of climate or soil. Growth is slow, and planting should be done during the rainy season, from seeds or by grafting.

He who has never tried a *"jabuticaba kiss"*, stolen from a loaded tree, should hurry as the harvest, while abundant, does not last long. Men and women of all ages, animals, birds and insects of all types share the fruit eagerly.

A very long-living tree, the magnificent jabuticaba takes a long time to reach fruition, but when it does, it bears fruit regularly, and the older the tree the more productive it is.

Putting on a veritable spectacle of beauty and bounty, when in flower the tree becomes covered in small white, strongly perfumed blossoms. Later, during fruition, the amount of fruit is startling for someone unfamiliar with the plant. Thousands upon thousands of flowers, and later the fruit, sprout and develop tightly packed over the surfaces of the branches and even the trunk, right down to the ground. Upon these occasions, the jabuticaba tree is loaded with fruit in all stages of maturation, usually coloring the tree with shades from green and purple to almost black.

Some varieties of jabuticaba bear fruit with thin

red stripes, on others the fruit is olive-colored with dark stripes. The fruit is completely spherical and usually about the size of a marble, though there are varieties where the size can be closer to that of a large plum. In all cases though, the tough dark skin bursts readily when bitten into, releasing the juicy white pulp, usually pleasantly sweet and surrounding a maximum of four small seeds in each fruit.

There are many types of jabuticaba trees, a veritable collection ranging from 12 to 15 different varieties. Among these about half are very productive, while the remainder are less so. Of them all, the Sabará is the most cultivated and better known for its sweetness and flavor. The Paulista, a larger tree than the other varieties, is large in all respects, including its purple fruit and its production. The Rajada has large, greenish, very sweet fruit. The Ponhema is best for making jellies and sweets.

In fact, this is not a privilege exclusive to the Ponhema jabuticaba, as the juice of any of them, obtained by maceration and boiled with sugar, results in a splendid jelly, which may be used as a spread on bread or as a sauce with meat dishes. The fruit with their skins may be fermented to produce a home-made liqueur, which is highly appreciated in the interior of the country. In Goiás the semi-ripe skins, still slightly green, are used to make a preserve.

All the jabuticabas are native to Brazil and today may still be found growing spontaneously throughout most of the country. They occur more frequently, however, in Minas Gerais, Espírito Santo, Rio de Janeiro, São Paulo and Paraná. Originally the trees probably grew in areas along the banks of rivers and streams in the southeastern region, forming thick copses and forests, and expanding naturally as much as they do under cultivation.

Ever since man learned to cultivate jabuticabas and enjoy their fruit, the tree is a must in any orchard or large garden. On farms in southern Minas Gerais and in São Paulo, it was a common practice - and it would be good if it were to remain so - to maintain large orchards formed exclusively of different varieties of jabuticabas, with no commercial intentions, just to provide fruit for the fortunate families and surrounding neighborhoods.

In spite of all its qualities, its flavor and the abundance of fruit offered every season, the jabuticaba continues to be to this day a fruit tree almost exclusively for home orchards or small farms. That is, there are no truly commercial jabuticaba orchards. The two main factors that restrict its cultivation are first the costs and difficulties of harvesting in an orchard with many trees, and secondly the difficulty in keeping the fruit, as it must be picked when fully ripe and in begins to ferment practically right away.

Which is why someone, who has been a child, like the poet, goes wild when he finds a jabuticaba tree loaded with fruit, and knows that "*Jabuticabas have to be eaten up in the tree*"!

LIME

Scientific Name: *Citrus aurantifolia* Swing var. taiti
Botanical Family: Rutaceae

Origin: Asia.
Plant Features: A medium-sized tree, may grow up to 4m in height, a straight trunk and a dense, rounded crown. Flowers have pink buds and yellowish-white petals, in clusters of 2 to 20 blossoms.
Fruit: Rounded shape, with a smooth or slightly rough rind, colored light green. The whitish succulent pulp surrounds a clear liquid. This variety has no seeds.
Cultivation: Propagated by grafting, the plant prefers climates and soil similar to that preferred by the orange. Picking may be done throughout the year, depending on the time of planting, most production is from April to September.

The origins of fruits of the *Citrus* genera have been lost in time, along with the origins of mankind. It is only known that the majority of citrus fruit originated in regions somewhere between India and the southeastern Himalayas where, still to be found growing wild, are varieties of lime, citron, lemon, grapefruit, sweet and sour orange trees, along with other acclimatized or native acid fruit trees.

The history of their introduction to the European continent is involved in a tangle of dates and facts. Jorge Sintes Pros says that the citron was known to the Romans before the Christian era, and that the bitter orange and the lemon were probably brought to the Mediterranean when, a century later, the Romans discovered a direct route between Europe and the East Indies via the Red Sea. The sweet orange, the same author claims, would have been introduced into Europe by the Portuguese only during the century of the great discoveries. Other authors push this date back to the 7th and 9th centuries, crediting the Moslems with the introduction of all these fruits into Europe at the time when they occupied large areas of the continent.

What is certain is that in America, citrus fruit arrived in the 16th century together with the European Conquistadors. But as much in Europe as in America, it was not until the late 19th century that cultivation and marketing of the various varieties took hold.

The lime is a fruit with prodigious virtues. In addition to being a powerful source of vitamin C, popular medicine attributes the lime with a number of curative properties, among them of acting as a natural antibiotic and as a regulator of the cholesterol count in one's system.

In cooking, it has countless applications, a few drops of lime juice are enough to transmit its flavor to other food. For example: its juice is used as a seasoning in preparing and imparting flavor to fish, seafood and fowl; its rind in pieces or grated is also an aromatic seasoning or an essential raw material for deserts, preserves, puddings, candy, custards, fillings, meringues, sauces, etc.; like almost any fruit, limes may be stored in syrups or as preserves; their refreshing juice may be used to make beverages, cocktails and ices. An enormous list of delicacies.

Finally, a Brazilian invention *par excellence*, and the country's specialty to be remembered, is the famous *"caipirinha"*, a cocktail where pieces of lime are crushed with sugar and then covered with a generous amount of sugar-cane spirit.

In Brazil, limes are misnamed, as the group of fruit known as sour oranges by botanists or acid limes by horticulturists are all called lemons, and the most commonly produced lemons in Brazil are in reality limes (Citrus aurantifolia) and the so-called "Galician lemon", also a lime. The real lemons, grown and consumed for hundreds of years in Europe, with their thick yellow skin, are known in Brazil as Sicilian lemons, and are hard to find in the tropical regions of the world. They are basically cultivated in subtropical climates, whereas limes are better acclimatized to tropical and equatorial regions such as Brazil.

Thus, in an example of cultural adaptation in Brazil, lemons were calmly and simply replaced by sour oranges or acid limes, with their thinner, smooth and completely green rinds, which also became known as lemons. In fact, they both have similar properties and uses in medicine and cooking, the one replacing the other.

Brazil is today a great producer of the lime. It is a hybrid fruit, and its origin is unknown, although in Brazil it is called a "Tahiti lemon" as it was taken from the Tahiti islands to California around 1875. However according to the IBRAF (Brazilian Fruits Institute), commercial cultivation of the lime began much later, in the 1950's in Florida, but only captured the North American market in the 1980's.

In Brazil, although there are references to its existence in the country early in the 19th century, its cultivation only began toward the end of the 1960's. Ten years later, when the Gallician lemon plantations were decimated by a blight, cultivation of the lime finally enjoyed a strong impetus and grew to its current proportions.

Today, highly productive lime plantations are spread out all over the country, with the exception of a few cooler regions in the South, the state of São Paulo being the largest Brazilian producer. Practically all the South and Central American countries, as well as Mexico and Florida, produce limes, with Mexico and Brazil heading the list. This, according to IBRAF is because, besides the high internal consumption of the two countries, where the lime has completely replaced the lemon, part of this production supplies over 70% of the world's requirement of juice concentrates, as well as the essential oils extracted from the rind.

LICURI

Scientific Name: *Syagrus coronata* (Mart.) Becc.
Botanical Family: Palmae

Origin: Brazil - Atlantic forest.
Plant Features: A palm, up to 12m in height, with a trunk that retains the bases of the leaves attached to its surface in a spiral. Large leaves up to 3m in length are arranged in a spiral. Small, yellow flowers in clusters appear predominantly from May to August.
Fruit: Egg-shaped, orange in color, ripening from October to December. A yellow pulp and cream-colored seeds.
Cultivation: Requires a substrate rich in organic matter and shade. Propagates from seeds.

"In the granitic formation of Bahia, subject to long periods of drought, with no arable land, to be found over wide areas, are the palm trees that sustain domestic animals, which in turn support man. The goats in the scrub land eat the licuri fruit. Returning home, they ruminate the pulp of the fruit, discarding the hard seeds. Women and children then collect the harvest. They break the seeds open and remove the kernels, which they eat, use to feed their chickens or sell on the market to the oil industry. These palms attract and hold man in these inhospitable regions. Without them there would be vast semi-deserted or depopulated areas."

PALMEIRAS DO BRASIL (Palm Trees of Brazil)
Gregório Bondar (1964)

The licuri is one of the principal palm trees of the semi-arid region in northeastern Brazil, where it is native.

According to Afrânio Fernandes, the Indians called the rustic vegetation of the area by the name of "caatinga" - *caa* signifying forest and *tinga* meaning open or clear - to differentiate it from the dense, dark forests of the tropical environment. There, in country with a hostile climate, where long periods of drought alternate with rainy seasons that are often violent albeit more brief, is where the licuri palm grows.

As it is able to withstand long periods of drought and bear flowers and fruit for long periods of the year, the licuri is a fundamental provider of resources for the survival of man in the region, and is also used to feed cattle and other domestic animals.

In such a hostile environment, the licuri palm is one of the main sources of food for many wild animals. The blue macaw, for instance, is a precious species that makes use of the palm nuts. Living in the hinterlands of Bahia, this bird is threatened with extinction - only some 70 wild specimens are left - and its survival as a species is intimately connected with the presence of the licuri.

Although it manages to germinate and grow out of gravel and requires little water for its survival, the licuri, contrary to its hardy appearance, occurs naturally in good, fertile soils. According to Harri Lorenzi, the soils of the hinterland are good land, producing a large quantity of viable seeds, disseminated easily by wild animals.

The licuri is to be found in these types of soils, mainly from the state of Pernambuco to southern Bahia, via the interior, its boundary being the Jequitinhonha Valley. Author Gregório Bondar claims that one could define the licuri as native to Bahia.

With its elegant bearing, average height and ornamental appearance, the licuri is easily recognized due to the spiral patterns around its trunk formed by the stumps of former fronds. Like the carnaúba, but to a lesser extent, by scraping the licuri fronds - often work for children - a powder is obtained, which is transformed into wax. The dry palm fronds are also a raw material for the production of utilitarian and handicraft objects, such as brooms, hats, dusters, etc.

In addition to serving as an important food element for man and animals and furnishing a good edible oil, licuri fruit is often eaten daily as a tidbit. Sold in strings reminding one of a rosary, the small nuts are removed one by one, to be eaten at any time of the day, at bus stops, on the way to school, or during one's daily chores.

ORANGE

Scientific Name: *Citrus aurantium* L.
Botanical Family: Rutaceae

Origin: Asia.
Plant Features: A medium-sized tree, may grow up to 8m in height, trunk with a gray-brown bark and a dense, rounded crown. Leaves have a firm texture and rounded edges, giving off a characteristic aroma when crushed. The small white flowers are aromatic and attract bees.
Fruit: The format and coloring of the fruit vary from one variety to another. Frequently with an orange-colored rind enclosing a juicy pulp that may vary from pale yellow to red. Rounded, flat seeds are whitish-green. Fruition is throughout the year, more abundantly from April to September.
Cultivation: Planting should be done at the beginning of the rainy season. Preferred climates have temperatures of between 23 and 32° C. Resistance to cold varies with the variety. No requirements regarding soil, but deep soil preferred. Propagated by seeds or by grafting.

Ancient Greek legend tells us that oranges were the real golden apples so well guarded by the dragon with 100 heads in the Garden of the Hesperides. To obtain them, as part of his eleventh Labor, Hercules fought tirelessly.

This legend is at least an indication of the antiquity of the orange in man's life and culture. If that is not enough, Pio Corrêa tells us that the cultivation and use of oranges goes back to a period two thousand years before Christ, as shown in writings found in China.

It is believed that the orange, along with other fruit of the *Citrus* genera, originated in Asia, where man learned to cultivate it and from where it spread to conquer the world.

In spite of all this antiquity and the knowledge the Greeks had of its existence, the introduction of oranges, limes, citrons, lemons and grapefruit into Europe came a great deal later, there being no records of this fact earlier than the 15th century. Some authors, however, claim that the Arabs had introduced some species of citrus fruits into the Iberian and Italian peninsulas a long time before that. In the American continent, the land and its inhabitants only became acquainted with the delicious and vitamin-rich fruit after the arrival of Columbus.

It appears that all the many species and varieties of Citrus fruit existing in the world - among the oranges alone there are close to 2 thousand different varieties, of which less than 100 are cultivated on a large scale - originated from not more than 10 or 12 wild species crossed among one another, transformed, selected, crossed again and improved over centuries and centuries. As a result of their ancient culture, the original wild forms of orange have never been found and have been lost in time.

Also as a consequence of this antiquity and the countless genetic changes they have gone through over the years, oranges as well as all other citrus fruits have been named and renamed many times in an attempt to structure a classification suitable to both science and commerce. For this reason, in the case of oranges there is often no sense in the nomenclature given to this or that fruit, particularly as regards their popular names, especially as sometimes the same variety may present differences in coloring and flavor due to climate and soil differences between one region and another.

Today, a large part of the tropical and subtropical latitudes of the world have been transformed into a veritable belt producing citrus fruits, making the orange one of the fruit most cultivated in the world.

In Brazil, orange production developed rapidly in the 1960's, when an unusually severe frost destroyed much of the orange plantations in the USA. As the US is the world's largest consumer of oranges, they had to import them, which encouraged countries like Brazil to invest in the culture. And it worked. The São Paulo producers were the first to move into the market. Over the last 20 years, the orange plantations have changed the landscape in many productive regions in the state of São Paulo, today Brazil's major producer of oranges. Enormous tracts of land, especially in the neighborhood of the cities of Limeira, Bebedouro and Araraquara, have been planted in oranges, and, when the trees are in bloom, their perfume turns the air in the area heady with the sweet scent.

Besides São Paulo, other Brazilian states also produce considerable quantities of citrus fruits, nota-

"- I saw everything - said the Viscount. The Hesperides live in a marvelous palace in the center of a garden. Right in front, there is a tree loaded with fruit the size of oranges, colored golden yellow. That must be what we are looking for.
- Why did you not bring a fruit? Were there none on the ground? - asked Emilia.
Pedrinho laughed
- How ingenuous! Do you think it possible that golden fruits roll around on the ground, like the oranges in our own orchard?"

THE TWELVE LABORS OF HERCULES
Monteiro Lobato

bly Rio de Janeiro, Minas Gerais, Sergipe, Rio Grande do Sul, Paraná and Goiás. Vast orange plantations flourish under the highest standards of quality and productivity, sufficient to supply the total requirements of the domestic market and a large share of the international market. For such, the country's fruit growers rely on the work of important research centers specializing in the development, perfection and genetic improvement of many varieties of citrus fruit, as well as on the training, recycling and up-dating of professionals in the area.

In the city of Cordeirópolis, São Paulo, for example, there is the Sylvio Moreiro Citriculture Center, an organ connected to the Campinas Agronomy Institute and considered to be the most important center for citriculture publicizing, training and research in Latin America. There, from a collection initiated almost 70 years ago, roughly 1900 different types and varieties of *Citrus* from all over the world are grown, providing stock for the entire country and much of Latin America.

The volume of business and money involved in the orange industry today is enormous, and, for this reason, the species and varieties grown are those that have the most market value. Among the various existing varieties of oranges, those most cultivated, best known and most consumed also have different uses.

Thus, over 80% of Brazilian commercial orange orchards produce fruit optimized for domestic or industrial processing of juice, this being a rapidly expanding market in the country. Varieties such as Pera, Seleta, Valência and Natal may be found throughout the year in markets, supermarkets and fruit shops all over the country. These juice oranges, also eaten *in natura*, combine deliciously with some salted dishes. As an example, one typical Brazilian meal is the *feijoada*, a combination of several meats, black beans, rice and kale, where the orange, cut into sections or sliced, forms a prominent part to *"cut the grease"* and attenuate the effect of the red pepper sauce.

Sour oranges are also frequently used in cooking at an international level for preparing sauces for cooking and to accompany meats, birds and fish, such as the famous *canard aux oranges*.

Then there are the Bahia and Bahianinha oranges, species developed in Brazil, more specifically in the state of Bahia, and today growing in various countries around the world. Like all the navel oranges, these are sweeter and more suited to being eaten *in natura.* In heavy demand in European markets, the Bahia is there considered to be the best of table oranges, as its firm consistency makes it ideal for elegant and sophisticated consumption with a knife and fork.

Below the equator, however, the qualities of various oranges are enjoyed in a somewhat less sophisticated manner at any hour of the day - whether for breakfast, as a dessert with lunch, with an afternoon snack, at night, or simply to restore energy, it is always the right time to enjoy the sweet, refreshing vitamin-rich juice of an orange. In the summer, children enjoy playing in Grandma's back yard and eating mountains of oranges patiently peeled by Grandpa, to then be thoroughly hosed down.

The sweet, mild-tasting "lime-orange", as it is misnamed in Brazil, has less commercial value and therefore is often hard to find. Less acid, very sweet and flavorsome, especially when picked from older trees cultivated with care in well-kept orchards, its juice is indicated for infants and all those suffering from digestive problems.

Also worthy of note are the oranges suitable for making delicious preserves, crystallized fruit, candied peel, jams, marmalades, jellies, creams, puddings, mousses, cakes, biscuits, tarts, coverings and any other delicacies that an ingenious cook may invent. For sweets, the more sour types of orange are usually preferred, such as the bitter or Seville orange, where the pulp and the rind may both be utilized, in some cases only the rind, which is patiently peeled with a special knife.

Also used in confectionery is the essence of the orange flower, an important aromatic. In Arab cooking, for instance, practically all sweets and syrups are perfumed with delicate flower essences.

A tasty, refreshing, nourishing, vitamin-rich, diuretic, cleansing fruit: *"praise for the orange"*, as Lúcia C. Santos states, *"is self-evident from the awesome consumption the fruit enjoys around the world".*

MACADAMIA

Scientific Name: *Macadamia integrifolia* Maid. & Bet.
Botanical Family: Proteaceae

Origin: Australia.
Plant Features: A tree, from 4 to 15m in height. Smooth leaves with wavy edges, colored grayish when young. White flowers appear twice a year.
Fruit: Round, up to 5cm in diameter, a reddish skin with a greenish cream-colored pulp, meaty, surrounding an edible kernel. Fruition is from February to May.
Cultivation: Prefers a mild, damp climate. Ideal for coastal plains. Cannot withstand frost. Prefers fertile, deep soils. Propagates from the seed, may be planted all the year round.

"When I came to work here, nothing surprised me, except for this nut with such a hard shell that it reminded me of my childhood, breaking palm nuts with a brick.
I cannot forget the macadamia. I found the nut very tasty and decided it would be successful in Brazil, where it has a fabulous future as a crop, and because of its economic importance. You should see the tree in bloom! It leaves a crazy aroma in the air, the bees seem to get drunk.
In the future we are going to have a large production of macadamia honey !"

Luiz Marino Netto
Companhia Dierberger, Limeira - SP

The fruit of the macadamia has an attractive, spherical shape, terminating in a small almost human-looking nipple. Green on the outside, the thin, meaty carpel of the fruit surrounds a nut which becomes exposed when the skin bursts. The still-closed nut falls by itself out of the tree. Within this nut, protected by a thick hard shell that must be broken mechanically, is the kernel.

The kernel or seed of the macadamia is edible, and may be eaten raw or cooked after drying, and used for confectionery, cakes and sweets as a replacement for other types of nut. It is most appreciated roasted, as a tidbit, and half of its worldwide production is utilized in this manner. The macadamia nut, as it is simply called, is highly nutritious, with a high fat content varying between 70 and 80% of its total weight. For this reason, extraction of its oil is extremely profitable and the quality obtained is comparable to olive oil.

The macadamia is a rustic tree of Australian origin, more specifically from the provinces of New South Wales and Queensland, where it was found in dense natural forests. Today it is grown especially in Australia and in Hawaii, where it was introduced at the end of last century, and to a lesser extent in Africa, Central America and California, where it arrived from Hawaii after the Second World War.

In South America, the plant acclimatized easily in the vast area from southern Bahia in Brazil as far south as Uruguay. The first macadamia trees to be planted in Brazil came from California to one of the Dierberger nursery gardens in Limeira, São Paulo, early in the 1930's, where production of seedlings for sale was initiated. Its cultivation, however, was destined basically as an ornamental in domestic orchards. In fact, the macadamia is a very ornamental tree, decorating any garden with class and elegance.

In 1948, technicians of the Campinas Agronomy Institute began experimenting with the adaptation of the plant for commercial cultivation in this country's climate. Early in the 1960's, Dierberger, in a new initiative, imported from Hawaii the seeds of one of the many existing varieties of macadamia, to improve the varieties to be cultivated in this country.

By the end of the 1980's, and particularly by the early 1990's, macadamia cultivation increased considerably. By the end of this century, it is estimated that the majority of the trees planted in Brazil during that period will have reached maturity and begun producing heavily, as the macadamia tree may take between 12 to 15 years to reach maximum productivity. In compensation, from then on it doesn't stop, in fact one of its main characteristics is its longevity - it is known that there are plants over one hundred years old that are still quite productive.

The macadamia nut is becoming increasingly appreciated on the international market, and business deals involving the macadamia industry turn over many millions of dollars annually. It is currently one of the most profitable crops in existence. Although most of the Brazilian production is directed to the domestic market, Brazil may well be on the way to becoming the world's largest producer and exporter of macadamias.

Today, there are great plantations of macadamia in the states of São Paulo, Minas Gerais, Espírito Santo and Bahia, many of them established along with other crops, such as coffee, oranges, lemons, tangerines and passion fruit.

MAÇARANDUBA

Scientific Name: *Pouteria ramiflora* (Mart.) Radlk.
Botanical Family: Sapotaceae

Origin: Brazil - Amazon and Mid-Southern regions.
Plant Features: A tree, up to 30m in height with a straight trunk and a milky latex sap. Light-green leaves vary in shape and size. The small greenish-cream flowers appear from August to October.
Fruit: Green, egg-shaped fruit with a white pulp containing a brown seed. A favorite of bats, fruition is from January to February.
Cultivation: Prefers sandy soils, and grows from seeds, developing at a moderate rate.

Maçaranduba or maçarandiba, possibly a corruption of moçarandiba, which means "tree that makes you slip or slide", according to professor Pirajá da Silva commenting on the text by Gabriel Soares de Souza.

It would appear that maçaranduba trees - which are common in the American continent, especially in Brazil - have been known to the Indians for a long time. It is believed that they are native to South America. Many species with this name are to be found in the state of Pará and all the Amazonian region, as well as in the country's mid-southern forests and the Atlantic forests. The majority of them belong to the *Sapotaceae* family, as do the abiu, caimito and sapotes.

The maçaranduba is a large forest tree, often reaching a height of 40 meters. Among its main features, besides the excellent quality of its wood, are its tasty fruit and a latex which is also edible.

The only disadvantage of the maçaranduba is its height and the fact that it grows almost exclusively in forests. For this reason, to pick its fruit, as Paulo Cavalcante says, is a thankless task, as *"during fruition time, the first to arrive at the banquet are the wild animals - especially bats - in reality their legitimate owners".*

"The maçaranduba is a royal tree... Suffice it to say here that the fruit is the color and size of the fruit of the strawberry tree, with a tough skin and two seeds, which are discarded with the skin, the rest being eaten, as it is very tasty and sweet; and whoever eats a lot of this fruit, with the same name as the tree, has his mustache made sticky from the juice which is very sweet and sticky; and for the Indians to pick this fruit, they cut the tree down, as they do with all the tall ones. They grow along the sea coast or in sight of it."

NOTÍCIA DO BRASIL
Gabriel Soares de Sousa (1587)

"In this American land, for Man's pleasure and health, God created a fruit that is sunlight transformed into juice, pulp and sweetness; because there is no better sensation on the palate than a piece of fresh papaya, when it melts in the mouth and slides down one's throat. God also made it well, as, in His laboratory of infinite wisdom He also made it into the most perfect natural medicine that, in the form of fruit, brings benefits to human beings."

FRUTAS DE AMÉRICA
Clara Inés Olaya

PAPAYA

Scientific Name: *Carica papaya* L.
Botanical Family: Caricaceae

Origin: Tropical America.
Plant Features: A tree with a green trunk up to 8m in height. Large indented leaves. White or yellow flowers. Contains a milky latex sap.
Fruit: Elongated, smooth, variable size. A meaty pulp, colored orange-red with numerous black seeds.

Cultivation: Requires a hot, damp climate, does not withstand the cold. The soil should be fertilized or fertile, humous and well-drained. It is best propagated from seeds. The variety most consumed is the mamão-papaya, a small, much sweeter pear-shaped variety.

The origin of the papaya has been lost in time. What can be safely said is that, when the Europeans first landed on the American continent, the plant was already known and its fruit, leaves, latex and seeds were being used freely by the natives.

It is presumed that the papaya originated in tropical America where, according to Paulo Cavalcante, all the species described for the *Carica* genera are to be found, the majority spread around the sunny foothills of the Andes, in Colombia, Ecuador and Peru. From there, the fruit would have rapidly spread throughout the continent, due to the pace of its life cycle, the ease with which the plant propagates and multiplies and the rapidity with which it bears fruit.

The papaya grows rapidly and produces abundantly, flowering and bearing fruit often at the same time and all year round, preferably in hot, damp climates. Today, it can be found in practically all the world's tropical countries, and, together with the banana, is one of the most widely cultivated and consumed tropical fruits. In Brazil it is no different, the greatest production and the best quality of fruits coming from the state of Pará in the north, from the northeast, especially the irrigated areas in the São Francisco Valley, and from the states of Bahia, Pernambuco and Espírito Santo.

Some 30 or 40 years ago, however, the papaya was no more than a garden plant, although usually present in all domestic orchards and yards. A notable fact occurred little more than 20 years ago, when a new, smaller and sweeter variety of papaya was introduced into the market. Initially grown in the state of Pará and known in the southern regions of the country as *mamão-papaya*, *mamão-do-amazonas* or *mamãozinho*, the little papaya rapidly captured the market, with its always sweet flavor and its small size, ideal for individual consumption. Currently, this variety of papaya is being produced on a large scale, as much for export as for internal consumption.

The papaya is a small tree with a short life span, and the sexes are segregated to separate plants - male papayas and female papayas. The flowers on the male papaya are not transformed into fruit, but are important for pollinating the flowers of the female papaya. The male flowers stand out from the trunk of the tree, which makes it easy to identify the sex of a tree. Sometimes hermaphrodite flowers appear, giving origin to male papaya fruits, important in the fecundation process.

The papaya's skin is very delicate; when the fruit is unripe the skin is green. Gradually, as the fruit ripens, yellow or pale orange patches appear and spread until the green has entirely disappeared.

To open and eat a papaya is always a unique and surprising experience, since, although there are not many different varieties cultivated commercially, one papaya is never exactly the same as another in color, flavor and scent. The many seeds are connected by fibers to a cavity inside the pulp. They are black and shiny, and, although they are usually discarded, they are edible and reputed to have many medicinal uses, such as acting as a vermifuge or assisting the digestive process.

The papaya is usually picked while it is still green and is allowed to ripen off the plant. Often to speed up the ripening process and also to remove a faint bitter taste, it is customary to scratch the fruit longitudinally to release a white, sticky latex from the skin. This latex is rich in a substance called papain, which for a long time has been used as a home remedy for such varied functions as removing warts or tenderizing meat. Today it is known that papain is spread throughout the plant, especially in the green fruit, and gradually disappears as the fruit ripens. It is a substance with many medicinal and industrial uses.

In countries such as Sri Lanka, Tanzania and Uganda, there are large plantations of the fruit being grown for industrial purposes. According to Clara Inés Olaya, this latex, *"converted into a white powder is sent to European and North American laboratories, where it is refined, transformed, bottled, patented, labeled and marketed in remedies for treating gastric problems, powders for tenderizing meats, skin lotions, industrial products for clarifying beer and for softening leather and wool"*.

However, among all the many qualities of the papaya, its main one is still to excite the palate. Eaten ripe *in natura* or in the form of sweets, it is still a highly appreciated dessert. In Brazil, there are countless recipes using papaya as an ingredient; The ripe fruit may be used for a preserve, but green papaya preserves are unbeatable, either in a syrup or dry, cut into strips. A popular art form of the confectioner is the shape of these strips, producing patterns of flowers or spirals with the long green papaya "ribbons".

The center of the papaya tree's trunk may also be used for an exotic sweet which, according to Gilberto Freyre, is a crude sweet demonstrating the *"tellurian predilections"* of its appreciators. Finally, green papaya may also be used as a vegetable dish, cut into pieces and fried with seasonings, or even in soups or meat stews.

"From Pernambuco to Bahia, they brought the seed of a fruit called mamões; which have a very good aroma when they are grown on the trees, and allowed to ripen in the home... These seeds were planted in Bahia and they soon sprouted; the soil was so good that, by the first year, the trees were taller than a man and, in the second, began to bear fruit... The branches are the leaves themselves arranged like a palm tree; and the fruit grows from the trunk between the leaves."

NOTÍCIA DO BRASIL
Gabriel Soares de Sousa (1587)

MANGO

Scientific Name: *Mangifera indica* L.
Botanical Family: Anacardiaceae

Origin: Asia.
Plant Features: A large tree, may reach up to 30m in height with a dense, leafy crown. Leathery leaves, colored red when young, turn dark green. Small white or greenish flowers.
Fruit: Elongated, egg-shaped or rounded with a greenish skin speckled with black, yellow or pink spots when ripe. An edible, meaty, succulent yellow or orange-yellow pulp, fibrous in some varieties. A flat seed of variable size.
Cultivation: The tree grows in any soil, but adapts best in hot, rainy areas. Planting should ideally occur during the rainy season. Propagation is by seeds, grafts or layering.

"In the 16th century, when orchards were no more than a few rare fruit trees around rural homes, the Mogul Emperor Akbar the Great, who reigned in northern India from 1556 to 1605, planted one thousand mango trees near Darbhanga. This fact illustrates the value the mango had, so many centuries ago, in a country with a civilization going back several thousand years. Three centuries later, English fruit grower Charles Maries found some of those trees alive and flourishing. Which is an undeniable illustration of the extraordinary longevity of the mango tree."

FRUTICULTURA BRASILEIRA
Pimentel Gomes

The mango is a fruit native to Asia, more precisely to India, the southeast of the Asian continent and neighboring islands, and is undoubtedly one of the most widely consumed tropical fruit.

In spite of being cultivated in its original lands for over 4 thousand years, its introduction into other countries was very slow; mangos only began to travel the world with the discovery of sea trade routes between Europe and Asia early in the 16th century. Once again, it was the Portuguese who performed this task, taking mangos first to the east and west coasts of Africa and then to America.

It was not until the 1700's, according to Pimentel Gomes, that Brazil or rather Bahia, received the first seedlings of Indian mango trees. From there they went to Mexico in the 19th century, from where they arrived in Florida.

The delicious, excellent fruits of the mango are cultivated today in all the tropical and equatorial countries of the world. In Brazil, the plant has been widely disseminated. According to Pio Corrêa, the mango was the Asiatic tree to best adapt to the Brazilian climate, producing countless varieties, and becoming practically a constant feature in the northern and northeastern landscape of the country, and easily found under cultivation in Amazonia and the southeastern and mid-western regions.

Who has never seen a tree loaded with mangoes being raided by groups of youngsters, to the despair of the owner? What child in rural Brazil has not sated itself, sucking delicious mangoes down to the stone? Which of them afterwards did not wind up with their hands, face and chest completely covered by the fragrant juice? Which of them did not also struggle to clear their teeth of the fibers revealing their misdemeanor?

Different trees bearing different varieties of mangoes may be found everywhere, in farms, urban and rural orchards and gardens, on properties large and small, as well as in parks, squares and streets throughout the country. The city of Belém is an eloquent example of this bounty, having become famous for its mango trees. In another corner of the country, the mango trees in the city of Rio de Janeiro were so numerous that they gave their name to an entire suburb, a train station and even a samba school: the green and pink Estação Primeira de Mangueira.

While the mango is a "real fruit", with its meaty pulp, delicious, fragrant, consistent, filled with juices and sugars, the tree is also shady, long-living and bears fruit abundantly, all of which makes it a valuable tree to own.

The mango fruits are very attractive, varying in size, shape, flavor and colors. Externally, the mango may be green, green with black spots, yellow, orange, golden, rose pink or even violet. Inside, surrounded by a strong-tasting brightly-colored pulp that may be more or less meaty, more or less sweet, more or less fibrous, depending upon the quality of the fruit, is a large, flat, fiber-covered stone.

The great variation presented by mango fruit around the world is astonishing. Books list references of between 500 and 1000 existing varieties. In Brazil, mangoes also vary considerably; just among those most common and known to the general public, Pimentel Gomes cites and describes 36 varieties, all easy to find.

However, the varieties of mangoes cultivated in commercial orchards are more limited. These are basically strains obtained after a careful process of selection and improvement of the fruit, with a view to reducing the amount of fibers in the meaty pulp, and emphasizing the red and pink colors most appreciated for fruit destined for export.

In Brazil today, the following are the most popular varieties cultivated: Alphonso, Bourbon, Carlota, Coração de Boi, Espada, Golden Nuggets, Haden, Keitt, Kent, Rosa, Rubi, Sensation and Tommy Atkins. Some of these are native to and developed in the country, others originated abroad, mainly in the USA, where producers in the Florida region are great fruit developers.

The main producers of mangoes in Brazil are the states of São Paulo and Minas Gerais, which between them account for roughly 50% of the planted area and 25% of the total production. Next come the northern states, responsible for half of the national production, notably, Bahia, Pernambuco, Piauí and Ceará.

Brazil has obtained a growing share of the world's mango market, exporting large quantities to Europe and to the USA, but the main international trade is dominated by the Asian countries, especially India, with its gigantic volume of production.

Basically, due to its excellent texture and flavor, the mango is usually consumed *in natura*. However, its pulp may be used for many different recipes for delicious desserts, preserves, jellies, purées, ices and mousses.

In India, its native land, where the mango is highly important as a food component for the population, it has countless other uses, notably as "mango chutney", an ideal sauce for meats, curries and other Indian dishes, made from the pulp of the green fruit, cooked and heavily seasoned with spices and peppers.

"The apple tree grows spontaneously in many regions of western Asia. It can be found in a wild state scattered throughout woods and forests in almost all of Europe. Its culture is very old. The inhabitants of lakeside Swiss towns used to consume large quantities of apples. Cider has been made since ancient times. The plant was brought to Brazil from Europe and acclimatized in regions with cooler climates."

FRUTICULTURA BRASILEIRA (Brazilian Fruit Culture)
Pimentel Gomes

APPLE

Scientific Name: *Malus* sp.
Botanical Family: Rosaceae

Origin: Europe and Asia.
Plant Features: A tree, may reach up to 10m in height. The trunk has a brown, smooth bark and a rounded crown. White or pink aromatic flowers.
Fruit: Globular fruit, may present a strong depression at the point of insertion of the stalk. Red in color, it may have small greenish or yellowish spots. Fruition is from February to April.
Cultivation: Requires a mild climate and silicious clayey deep soil to develop. Varieties most cultivated are: gala, golden delicious and fuji, with varying textures and pulp flavors.

If the mangosteen is considered to be "the queen of tropical fruits", the apple in turn is "the queen of European fruits". The apple tree has been cultivated in the temperate regions for thousands of years. The antiquity of the apple's participation in the life of Man is a well known fact - tales involving apples may be found in the legends and myths of different civilizations, far removed from one another.

Worthy of mention is the apple in the Bible story of the Old Testament, telling of original sin in association with the fruit, used by the serpent to lead Adam and Eve to sin. Closer in time, there is the story of Snow White, one of the better-known fairy tales in Western culture, where an apple poisoned by the wicked step-mother, transformed into a witch, is an element involving witchcraft and unhappiness.

To this day, it is not known for sure when or where the apple originated, or which were the wild species to have originated the modern apple, the varieties of which are known today. For example it could have been *Malus sylvestris*, originating from Europe, *Malus prunifolia*, from Siberia and northern China, or *Malus pumila* coming from the Caucasus and part of Russia, or all of them together.

The apple is possibly the fruit with the greatest known number of varieties, which is estimated at somewhere between 5 and 20 thousand. Of these, some 3 to 4 thousand are cultivated on a larger or smaller scale in various regions of the world. It is certain that this number is growing daily with the improvement and development of new varieties in experimental centers.

As Eurico Teixeira says, the apple is "too well-known to be described". By varieties, the fruit can be classified and grouped by flavor, size, shape, appearance, the consistency of its pulp and its skin color, and by its various uses. Basically, they can be divided into three types: table apples, cooking apples and those suitable for making cider and/or vinegar. In spite of the countless varieties of apples in existence, one same tree may bear fruit with different uses, according to its classification. Thus, after passing through a first selection, those apples with a shape, color, size and appearance having a better commercial aspect are carefully packed for consumption *in natura*, while those that are slightly damaged, or otherwise do not meet specifications, are sent to be turned into sub-products, such as juice, cider, vinegar, alcohol, jellies, preserves, sweets, etc..

In Brazil, the first apple strains to arrive in the 1920's for establishing commercial plantations were planted in southern Minas Gerais, in the Maria da Fé area. There the trees flourished, due to the altitude and the climate which approximated that of temperate regions. However up until the 1960's the apple was a rare fruit in Brazil, not easy to find and kept for special occasions, or else reserved for feeding babies and convalescents.

In fact, with its high content of potassium and its capacity to provide quantities of fiber, the apple is indicated for maintaining good health, preventing heart disease and cholesterol in the blood, as well as for slimming diets.

Also, as it is highly prized for its delicate flavor and ease of digestion, the possibilities of consumption and utilization of the apple are very large, whetting the imaginations of cooks: *in natura*, raw, boiled or baked; with or without its skin; whole or in crunchy, succulent bite-sized chunks, in pieces, grated or as a purée; as an ingredient for desserts, jellies, preserves, cakes, tarts, crêpes and strudels; as an accompaniment for meats and in the preparation of many savory dishes; or as a beverage in juices, teas, wines and ciders; or even as a vinegar.

As of the mid-1970's, Brazil has become a large producer of apples, with a number of varieties having been developed in the country, such as the Rainha, Soberana, and the Brasil. Notable in this respect is the work of selection and adaptation of varieties conducted by the Campinas Agronomy Institute, which has developed strains able to produce satisfactorily under milder winter conditions. Also widely cultivated are varieties such as the Gala from New Zealand, Golden from Canada and the USA, Anna from Israel and the Japanese Fuji.

Very well adapted to the regional climates, the apples found on Brazilian markets come mainly from the south and southeast of the country, where the states of Santa Catarina, Rio Grande do Sul, São Paulo and Paraná are responsible for almost the entire volume of production. Today, Brazilian apples surpass, in quality and in the Brazilian customer's preference, the products with a traditional reputation such as the famous Argentine apples.

MANGOSTEEN

Scientific Name: *Garcinia mangostana* L.
Botanical Family: Guttiferae

Origin: Asia.
Plant Features: A tree, up to 10m in height with a conical crown. Large, hard, dark-green shiny leaves. Large, dark-red flowers.
Fruit: Spherical, from red to dark brown with yellow spots. A soft, succulent pulp, with a delicate very characteristic flavor, surrounds oily seeds.
Cultivation: The plant prefers hot, rainy climates, permeable deep soils, damp and well drained. Can be grown from seeds or through grafting.

Native to the tropical region of southeastern Asia and the majority of the islands of Indonesia, the mangosteen is considered by the inhabitants of those areas to be the tastiest fruit in the world - "the queen of tropical fruit". A veritable food for the gods, the mangosteen has been compared by some to nectar and ambrosia of Grecian Olympus, the drink and food of the Gods.

The mild, pleasant and unique flavor of the soft, succulent and aromatic pulp, surrounding five or six seeds of the mangosteen, is difficult to describe or compare. The fruit should be eaten fresh, *in natura*, as any additional element is liable to mask its delicate flavor, robbing it of its charm. All this aura surrounding the mangosteen, along with its exotic appearance, are associated with delicate and sophisticated delights, and, for this reason, it is one of the most expensive fruits on the international market.

The mangosteen arrived in Brazil, more precisely in Belém, Pará, in Amazonia, in the early 1940's. It was planted in what is now the Eastern Amazonia Agro-forestry Research Center of EMBRAPA, since when many experiments have been conducted and techniques developed to acclimatize the plant to the region and render its cultivation economically feasible. During this period the tree and its fruit adapted very well to the damp tropical conditions of Amazonia, at times producing better results than in its native country.

The mangosteen is a culture requiring little care and a relatively modest investment, as it can be planted along with other species having a quicker return, such as orange, lemon, tangerine or cacao. Once the long-living, short trees have grown to maturity, the only important requirement is during harvesting: the fruits should be picked, while they are still half green, at the exact time when their external color starts to turn from green to pink. Right after picking, the fruits must be refrigerated, as they spoil rapidly, and packaged in special cardboard boxes.

Although the mangosteen tree takes several years to begin bearing fruit - roughly 10 years when planted without a graft, or 4 years if it is grafted - cultivation of the "tastiest fruit in the world" has attracted many migrants of Japanese origin, especially those living in the neighborhood of Belém, many of whom being familiar with the fruit in their native land are attracted by the prospect of starting a lucrative production for export.

Another rapidly expanding mangosteen production center in Brazil is in the cacao region of southern Bahia, where the fruit was introduced a few years ago and has adapted quickly and easily to local conditions.

Today the majority of the mangosteen international trade comes from countries with tropical climates, especially in Central America and the Caribbean islands. However, in the early 1990's, Brazil began to export mangosteens to the USA, Canada and Europe, and is likely to expand its share of this market rapidly.

Passiflora maliformis L.

PASSION FRUIT

Scientific Name: *Passiflora* sp.
Botanical Family: Passifloraceae

Origin: Tropical regions.
Plant Features: A vigorous creeper with a frequently-grooved stem. Deeply indented leaves with serrated edges. Large, attractive flowers with colors varying from greenish-white , orange, red or purple, according to the species. Flowers from December to April.
Fruit: Usually spherical with a thick rind colored green, yellow, orange or with pale green spots, depending on the variety. Seeds are flattened, black, surrounded by an aril of a gelatinous texture, colored yellow or translucent. Fruition is all the year round but more intense from May to August.
Cultivation: The plant prefers hot, damp climates and is cultivated in all the tropical countries. Requires a humous, clayey, deep, fertile and well-drained soil.

One of the main characteristics of the passion fruit plant is its types of unusual flowers, of a beauty not to be found in any other vegetable species. The passion fruit, or maracujá, its fruit and its flowers, were well known and used in the Americas long before the arrival of the first Europeans, who soon became enchanted with their exuberance.

In the religious fervor of their conquest, the foreign missionaries saw in these flowers and fruits far more than beauty and fragrance. Clara Inés Olaya states that the religious saw in that complex and admirable formation a veritable present from God

Passiflora edulis Sims.

to illustrate his work of catechism, finding in its exotic forms and colors the perfect metaphor to explain *"to the infidel indians the truculent history of the passion of Christ"*.

So, they first associated the colors that nature had bestowed upon the beautiful flowers of the maracujá with the reds and the purples used in the Christian rituals of Holy Week. Besides the colors, the floral crown, completely filigreed, was transformed into the crown of thorns, with which Christ was crucified; the flower's three stigmas became the three nails with which he was hung upon the cross; the five others would represent Christ's five wounds; the tendrils were seen as the scourges that martyred him; and the round fruit were a representation of the world that Christ had come to redeem.

Ever since, the flowers as well as the fruit of the maracujá and other climbing plants of the same type, began to be called "passion fruit", after the Passion of Christ.

The seductive colors of the passion flower, together with the strange beauty of its forms and crowns, make of the maracujá an essentially ornamental plant. Ideally suited for covering fences and pergolas, the maracujás are climbers which, as they develop, hold onto other plants, fences or similar supports by means of their tendrils.

The flowers are also the object of great personal passions. Cases are related of

Passiflora alata Ait.

Passiflora quadrangularis L.

collectors who study them obsessively and cultivate them in the most adverse climates, maintaining greenhouses and complex equipment simply to enjoy the pleasures provided by the gentle and delicate opening of the flowers.

As if its esthetic qualities were not enough, many of the *Passiflora* vines also provide humanity with the famous and delicious fruits with countless food, culinary and medicinal uses.

Those drinking maracujá juice imbibe, among other things, a large amount of water-soluble vitamins, especially A and C, mineral salts and fibers, but the main prestige of the maracujá - possibly one of the best-known medicinal plants in Brazil - arises from its calming or sedative properties. In fact, from its fruit, leaves and seeds it is possible to extract large quantities of Passiflorina, a natural sedative.

In Brazil, culture of the passion fruit is progressing well. Today the country is an important producer and exporter of the fruit, notably from the state of Pará, accountable for over one-third of the national production. Next in importance comes the northeastern region, where the states of Bahia, Sergipe and Ceará together represent almost one-third of the total production, followed by the southeastern region with a quarter of the production, in which the state of São Paulo is the leader.

In some regions, however,

Passiflora edulis Sims.

"We began to talk about the excellent passion fruit, which, if not the queen, is the duchess of the fruits, because of the flower with which nature has ennoble and distinguished it, above all the fruits and flowers of the earth; the Creator has himself painted, on this mysterious flower, the lamentable tragedy of his Passion, the column, the scourges, the nails, the wounds, the crown, the blood, with so much perfection and vividness that, for this reason, the flower is called the passionflower."

FRUTAS DO BRASIL
Frei Antonio do Rosário (1702)

deforestation is harming the maracujá, as the destruction of the forests also causes the disappearance of the mamangava, a species of large bee, fundamental for pollinating the maracujá's passion flowers.

Today, many maracujás native to Brazil are under cultivation in other tropical countries such as Hawaii, Venezuela, South Africa and Australia, where they attain considerable economic importance.

The botanical genera *Passiflora* presents an extraordinary diversity of species, not counting the varieties and hybrids. Most of them may be found growing naturally throughout tropical America, where they are native. Only a small number have originated in Asia or Australia.

According to Eurico Teixeira, this abundance can only be explained due to the facility with which the flowers of the creepers produce hybrids and crossbreeds, and for this reason, "it is natural that new species should arise spontaneously and indefinitely".

Among so many species, not all produce fruits that are edible, and only a small number attain prominence in national and international fruit markets. The best known, and with the most commercial application, are basically two: the yellow and the purple maracujá, both varieties of a single species, almost perfectly spherical in shape, and the sweet maracujá, looking rather like a small papaya.

The yellow or sour maracujá (*Passiflora edulis f. flavicarpa*) is the one with which the Brazilians are most familiar, as its fruits are available all the year round, especially in the northern and northeastern regions of the country. This passion fruit, as yellow as the sun it is so fond of, contains a large number of small brown seeds. After its skin has become completely wrinkled is the best time to savor its succulent, sourish pulp. The yellow maracujá is the variety with the best productivity and is ideal for sweets, jellies, cocktails, juices, beverages and ices, sweetened to taste.

The purple passion fruit or granadilla (*Passiflora edulis*) is also spherical but smaller than the yellow variety. Reputedly less acid than the other, the purple granadilla is delicious for consumption *in natura*. The plant prefers subtropical climates, such as those of South Africa, Australia and southern Brazil.

Research and experiments conducted recently have managed to increase the sweetness of passion fruit destined for the industrial production of bottled juice, also increasing the weight and yield of the fruit and rendering it more resistant to disease.

The sweet passion fruit (*Passiflora alata*) in its turn, although native to Brazil, is a variety less known, planted and consumed by Brazilians. Its pulp is sweet, with a powerful, pleasant aroma, possibly even too sweet to be processed into juice, so the fruit is consumed almost exclusively *in natura*.

Besides the species of maracujá usually cultivated, there are hundreds of other wild species to be found in highly diversified environments, climates and soils. Some, unknown on international markets and in large cities, are cultivated exclusively in private orchards and experimental gardens, presenting an immense variety of delicate flavors and fragrances.

Notable in Brazil is, among others, the exceptional maracujá-açu, maracujá-melão or maracujá-mamão (*Passiflora quadrangularis*), also known as the *badea* or *granadilla gigante* (giant granadilla) in Spanish-speaking America. Native to the north of Brazil, the maracujá-açu does not thrive outside the Amazonian region. Its main characteristic is the size of its elongated, large fruit, which may weigh as much as 3 kilos. The skin is a pale greenish yellow, the pulp firm and thicker than that of the common varieties, and all of the fruit may be eaten, except for the thin skin. For this reason, the maracujá-açu is ideal for manufacturing sweets, mousses and ice creams.

Another species of passion fruit, notable for the manner of its fruition, as well as the beauty and ex-

Passiflora cocánea Aubl.

Passiflora amethystina Miran.

oticism of its flowers and fruit, is the *Passiflora maliformis*, also called the maracujá-maçã (apple maracujá) or maracujá-de-osso (bone maracujá). The flower of this passion fruit is large and strongly perfumed, and colored a pale creamy-green with tiny violet-colored stripes.

When pollinated, the three flaps separating the flower from its stem close over it and the fruit begins to grow inside, protected by a creamy-green colored cover. This cover has the important function of protecting the flower and fruit from the hordes of insects attracted by the perfume and sweetness. During fruition, the maracujá vine becomes heavy with little bags containing the small fruit.

The pulp, while very sweet, is difficult to get at since, as the skin of the fruit is extremely tough and hard, and cannot be broken with the bare hands or even with a sharp knife, it often being necessary to hit it hard with a hammer.

Also worthy of mention is the maracujá-da-serra or maracujá ametista (mountain or amethyst maracujá) (*Passiflora amethystina*), native to the Serra do Mar mountain range and other mountainous regions in the states of São Paulo, Rio de Janeiro and Santa Catarina. This species is exceptionally beautiful for its amethyst-colored flowers, the same color as the most Brazilian of semi-precious stones.

Another example of delicate beauty and the exoticism of its fruit and passion flowers native to Latin America is the vine bearing the maracujá-tomé-açu (*Passiflora coccinea*). The plant grows over an extensive region from the Guianas down to the Brazilian state of Rio Grande do Sul, and is distinguished by its scarlet-colored flowers, with the filaments ranging in color from purple to rose-pink or white. Also, its small oval fruit, colored orange with pale green stripes, resembles a miniature melon.

MARAJÁ

Scientific Name: *Pyrenoglyphis maraja* (Mart.) Burr.
Botanical Family: Palmae

Origin: Amazon region.
Plant Features: A palm with a long thin trunk, may reach a height of 8m.
Leaves are armed with spines along the parts close to the trunk. Yellow flowers form large clusters.
Fruit: Globular or elipsoid, a smooth skin and black-to-purple color.
A thin, mucilagenous yellow pulp surrounds a hard dark seed. Fruition is from March to June.
Cultivation: A wild species, abundant in flooded areas, river banks and backwaters.

In Amazonia, it is not easy to penetrate the forest, paddling small canoes through endless waterways and backwaters, without finding oneself amongst palm trees such as marajás, açaís and bacabas.

In Brazil, several small palm trees of amazonic origin are traditionally known as marajás, appearing frequently in swampy ground, along the region's river banks, lagoons and backwaters, where they are to be found growing together in large groups.

Both the palm tree known simply as a marajá or another similar one called marajá-açu are very common on the island of Marajó, on land that is under water for much of the year. Thus, abundant, spontaneous and native, the marajás are palm trees liv-

ing preferably in the best part of the forest, with plenty of shade and fresh water.

The fruits of the marajá usually have a hard blue-black skin enclosing a white pulp, which is edible *in natura*, more or less juicy, yellowish or reddish, tasty or tangy, depending upon the species. This pulp produces a wine-like liquid or a vinegar of a variable quality.

The marajá appears to be a fragile plant with a very slender trunk that might break in the wind. However, the wood is very tough and extremely hard, much used for making such diverse objects as walking sticks and fence posts. The Amazonian Indians, for example, make the most of these qualities, using the wood to make weapons, arrow tips and lances. Due to its great resistance to humidity, it is also used for building underwater fish traps.

MARMELADA-NATIVA

Scientific Name: *Alibertia edulis* (L. Rich.) A. Rich.
Botanical Family: Rubiaceae

Origin: Amazon region, in copses and fields.
Plant Features: A tree, up to 4m in height with a crown 3m in diameter. Shiny, bright green leaves. White flowers.
Fruit: Globular, with a smooth skin, black when ripe. An edible black pulp encloses numerous seeds. Fruition is throughout the year, but is more bountiful from September to November.
Cultivation: A wild species, with no information as to its cultivation.

The marmelada-nativa is a wild fruit quite common in the Amazon region and the Brazilian savanna areas. In Amazonia, where it is also known as puruí, it grows in copses as well as in open fields.

The plant is a small tree, bearing fruit the size of a pomegranate which, when ripe, its skin and its pulp are colored black. Sweet-tasting, the marmelada-nativa may be eaten *in natura* or used to prepare sweets, especially jellies. Each fruit contains from 10 to 30 seeds, which the local population in poorer regions roast and grind as a coffee substitute.

Marmelada, marmeladinha, marmelada-nativa, marmelada-do-campo, marmelada-macho, marmelada-de-cachorro, marmelada-de-bezerro; contrary to what it would seem, the plant has no botanical resemblance to the "marmelo" or quince. They do say however that the flavor of the various Brazilian marmeladas reminds one of the tasty quince sweet or jam.

According to Eurico Teixeira, father Aires de Casal, writing about the fruit trees of Goiás state, made reference to the various marmelada trees he found there. He singled out one species, the fruits of which were as large as and shaped like oranges, but which turned black when ripe. He stated that the pulp of the fruit had a flavor that was "sweet-sour and refreshing", reminding one of quince jam (marmelada), so they were given the same name.

This description matches the marmelada-nativa of the savannas, which, it appears, has been known and used since the days of the first explorers.

WATERMELON

Scientific Name: *Citrullus vulgaris* Schrad.
Botanical Family: Cucurbitaceae

Origin: Africa.
Plant Features: A ramified creeper with oval leaves, divided into 3 lobes and tendrils attached to the stalk. Small greenish-yellow flowers.
Fruit: Round or elongated, with a smooth rind, green or striped with yellow. An abundant pulp varying from pale pink, yellowish, red or purple, with red or black seeds.
Cultivation: From the seed in August in regions with a harsh winter, and at any time of the year in milder climates. The plant prefers deep silicon-clayey soils, rich in organic matter. The fruit can be gathered roughly 100 days after planting.

The watermelon is an unusually large fruit, possibly one of the largest of all the existing fruits. To get some idea of this, under normal conditions of production, its weight may vary from 10 to 20 kg of pure water! Probably originated in North Africa, it is believed that the watermelon has been known to and cultivated by man for over 4 thousand years. The plant requires hot or temperate-hot climates, and today it is to be found in all the tropical regions of the world.

In Brazil, the fruit may be planted anywhere, and it is! In volume and in economic value, the watermelon occupies one of the first ten places in the list of vegetables sold on the domestic market. The main watermelon production centers are in the south, with the states of Rio Grande do Sul and São Paulo supplying almost half the production - and in the northeast, where the irrigated areas of the São Francisco Valley account for about one quarter of the total production. Some of this is exported, especially to other countries in South America.

Together with the melon, the watermelon belongs to the *Cucurbitaceae* family, as do the cucumber, the pumpkin and the chayote. Among their many different varieties, although there is one in which the pulp is yellow, watermelons usually have a pulp coloring which varies from pale pink to a strong red. No matter which variety, the watermelon is completely and randomly dotted with small, smooth flat black seeds dispersed throughout the pulp. This pulp is very watery, more so than that of the melon. Pure water! For this reason, the watermelon is one of the most refreshing fruits in existence, more so than the melons, as it is also less indigestible.

It is a fact, that in times of extreme heat, under the noonday sun of a tropical summer, there is nothing like a good chunk of fresh watermelon to replace the liquids lost to perspiration. Undoubtedly this is the best time to enjoy the fruit!

In the cities in the northeast of Brazil and in some of the warmer cities of the southeast, it is common to see vendors selling chunks of watermelon and other refreshing fruit to cool their customers. Due to its size, a whole watermelon is often too much for one single family, so they are usually sold in markets already cut into pieces, quarters or halves.

The abundant juice and firm pulp of the watermelon are very sweet, and usually the riper the fruit, the sweeter it is. Sometimes, however, the fruit is tasteless and insipid, fit only for feeding livestock, who never tire of it. It is not easy to determine whether a watermelon is ripe or not, as its thick, hard rind tells one nothing. One method is to knock on the rind with the knuckles; if the fruit is still unripe the sound is slightly metallic. On the other hand, the duller the sound coming from the fruit, the riper it should be.

MELON

Scientific Name: Cucumis melo L.
Botanical Family: Cucurbitaceae

Origin: Asia and Africa.
Plant Features: A creeper with an angulate stem, large leaves divided into 3 to 7 lobes with a velvety texture. Yellow flowers.
Fruit: Large and round, with a meaty, green, yellow or salmon-colored pulp. Abundant seeds in the center of the fruit.
Cultivation: Hot, dry climates, with ideal temperatures between 25 and 32 °C. Dry soil, rich in organic matter. Grown from the seed. The fruit may be gathered roughly 100 days after planting.

The melon and all its varieties belong, as does the watermelon, to the *Cucurbitaceae* family and are species very closely related to some important legumes, such as the cucumber, the pumpkin and the chayote.

The melon probably originated in the central regions of Asia and Africa, from where it would have been taken to Europe at the beginning of the Christian era. Known and cultivated for many centuries, the fruit was much appreciated by the Romans. It is thought that both melons and watermelons were also cultivated by the Egyptians.

There are many different types and varieties of melon. They vary in size, shape and skin color, and

in the color and flavor of the pulp, but they have one common feature: their pulp is juicy, smooth and delicately sweet.

Composed basically of water with very few calories, the melon is a very refreshing fruit and may be indicated for various types of diets, for slimming, or for rehabilitating convalescents, for example.

For a long time, melons were considered in Brazil to be a delicacy found only in private vegetable gardens, where, with their fine flavor and rarity, they were exclusively eaten *in natura* and in tasty fruit salads.

According to confectioner Lúcia C. Santos, in 1950 it constituted - and still does - *"almost a crime to transform a melon into a sweet, as the heady aroma and exquisite flavor would be lost"*.

To this day, the melon is basically destined for consumption *in natura* and is present in international cuisine to be served not just as a dessert. The fruit may be garnished with ham or other smoked meats as a sophisticated entrée for summer meals and an accompaniment to other savory dishes.

Today melons are being cultivated throughout the tropical and subtropical regions of the world, and some varieties even grow in temperate regions under extremely cold conditions. Production and sale of the fruit during the second half of this century have become big business, involving large capital investments, and Brazil, keeping up with the times, has changed from being an importer to becoming a large producer and exporter of melons to European and North American markets; some 20% of the country's production is exported.

The northeastern region is currently one of the great fruit-growing areas of Brazil. In the case of melons, the states of Rio Grande do Norte and Ceará produce almost two-thirds of total output, and the states of Pernambuco and Bahia almost one-third.

As from the 1980's, there has been a great change in the picture, especially in the Açu River and São Francisco River valleys, in the municipalities of Petrolina (PE) and Juazeiro (BA). With the installation of large agricultural projects, provided with modern irrigation systems and specialized professionals working to improve the varieties under cultivation, veritable gigantic oases have sprung up, where the fruit is grown under very strict quality conditions. Even the bees that pollinate the melon flowers receive special care. The result is sweet, healthy and attractive fruit that leaves packed and ready for export or to supply markets all over Brazil.

The many varieties of melon existing in the world are usually grouped, from a commercial point of view, into separate categories, according to the sweetness and aroma of their pulp. Weight and size are also factors determining the classification of the fruit for markets, especially when they are to be exported.

International fruit markets are today demanding and determine the standards of quality of the products due to reach the consumer's table. For example, besides being sweet, a good melon must have a juicy and firm pulp, pale green coloring, ranging to orange around the edges. The many seeds, surrounded by a kind of fiber, should fill the fruit's central cavity entirely , and be firmly held in place. A typical external color is a not too deep yellow, reached when the fruit begins to ripen, which should be homogenous, with no spots, scratches or sunburnt areas.

MAMA-CADELA

Scientific Name: *Brosimum gaudichaudii* Tréc.
Botanical Family: Moraceae

Origin: Brazilian scrub land regions.

Plant Features: A tree with a crooked trunk that may reach 8m in height. A gray, thick bark. Rigid leaves with whitish hairs on the lower surface. Flowers clustered around a stalk in a rounded shape, attached to the insertion point of the leaves on the branches. Flowering varies, mainly from August to November.

Fruit: The small fruits are attached to a meaty stalk forming a globular orange-colored structure when ripe. Fruition is from September to November.

Cultivation: Occurs spontaneously in Brazil, no information on cultivation is available.

The mama-cadela is a spontaneous species found throughout the savannas of central Brazil, where it grows as a large tree.

Also called the mamica-de-cadela, this fruit is prodigious in the associations and memories it recalls, beginning with is its name (bitch teats), only one of its various folk names.

For some authors, contrary to what one would expect, the name mama-cadela does not refer to the similarity in shape the fruit might have to the teats of a female dog, inasmuch as the shape is very different. The plant and the animal are similar in the way that the mama-cadela fruits hang side by side, almost horizontally, along the long branches, really reminding one of the positioning of the tits on the belly of a female dog. To increase this similarity, small animals of the savanna may frequently be seen rising up on their rear paws to suck the juice of the mama-cadela, a scene much like that of feeding pups.

When ripe, the fruit of the mama-cadela turns from dark green to a dark, almost orange yellow and contains a good deal of sweet juice.

Although it is small and has a sticky pulp, this fruit is much sought after, especially by children. It is sucked, savored, and chewed like a gum until only a tasteless residue remains. Pimentel Gomes goes as far as to say that the mama-cadela is a kind of "natural chewing gum." The consistency and appearance of the residue is much like that of cotton or sponge or cotton waste and, for this reason, *algodãozinho* and *algodão doce* (little cotton ball and sweet cotton) are among the fruit's nicknames.

MONGUBA

Scientific Name: *Pachira aquatica* Aubl.
Botanical Family: Bombacaceae

Origin: Brazil - Amazon region.

Plant Features: A bushy tree, up to 14m in height, with a slightly curved trunk and many branches. Compound leaves with 3 to 5 dark green leaflets. Large yellowish-white or rosy-white solitary flowers up to 20cm in length appear between September and November.

Fruit: An elongated or egg-shaped fruit with a velvety, reddish rust-colored skin. Large, rust-colored seeds contain an edible white kernel. Fruition is from April to June.

Cultivation: The plant adapts best when cultivated on the banks of rivers, lakes and flooded land, but can also thrive on drier soil. Also used as an ornamental.

The monguba, mamorana, or castanheiro-do-maranhão tree originates in the Amazon forest, where it may frequently be found amongst riverine forest growth. Well-known in all the Amazon, especially in the great river's estuary, the monguba is spread throughout the region and reaches the lands of the state of Maranhão.

Spontaneously, the tree grows in damp areas along

the banks of rivers and lagoons, or in swamp land, hence the aquatica in its scientific name. Nonetheless, the monguba adapts easily to quite diverse soil and climate conditions.

In general, it varies in size and its lush foliage forms a dense, rounded crown. Owing to these characteristics and to the exotic beauty of its large yellow red-tipped flowers, it is admired for its ornamental function. Consequently, it is used to line the streets of Recife, Rio de Janeiro, and Belo Horizonte, thus proving its adaptability and ability to thrive even in dry soil.

Although it is a well-known species, adaptable to cultivation, and has delicious fruit that can be used in various ways, the monguba is little utilized by Brazilians; it is not recognized as being important in financial or commercial terms, which is a total misconception.

The beautiful mongubas produce large amounts of fruit that all the fauna dispute. They eat the seeds. Belonging to the paineira, or silk-cotton tree family, the monguba's seeds, kept in large, long velvety reddish-brown pods, are further enveloped in a silky, white cotton.

The nuts are edible and may be eaten raw, roasted, fried, cooked with salt, or toasted. In the Guyanas and Antilles, to which the tree was transported from Brazil, as Pio Corrêa tells us, the shelled monguba nut is transformed into a nutritive, much appreciated flour that can be ground, toasted and made into a beverage that serves as a substitute for coffee, chocolate and milk.

"The strawberries are presented in large quantities, in a silver basket or crystal bowl, decorated with green leaves. In either case, the berries must first be carefully washed and their stems cut off. Once in the dish, confectioner's sugar is sprinkled on the berries, which are then eaten one by one with a fork or spoon. The fruit should not be mashed in the plate. In some parts of the world, it is customary to serve strawberries simply washed, without cutting the stems, so that they may be eaten, one by one, with the hands. We, however, avoid this eccentricity."

BOAS MANEIRAS - Manual de Civilidade
Carmen D'Ávila (1936)

STRAWBERRY

Scientific Name: *Fragaria vesca* L.
Botanical Family: Rosaceae

Origin: Europe.
Plant Features: A small creeping plant. Leaves composed of three leaflets. White or pink flowers.
Fruit: Small, grouped around a fleshy red stem when ripe. The small black hard spots, popularly known as the seeds, are actually the fruit of the strawberry. Fruition is from April to August.
Cultivation: The plant requires sunlight and dry siliceous soil, rich in organic matter. It propagates by means of scions, from the division of runners growing around the base of the parent plant. Many varieties are known.

In 1936, as the above quotation from Carmen D'Avila's work leads us to believe, strawberries were still considered exotic on this side of the world. In fact, large, sweet, succulent strawberries, like those found today for sale in street and other markets, were, for long, only an ageless dream for horticulturists worldwide.

The strawberry belongs to the *Rosaceae* family, that of apples, pears, quince, plums, dates, peaches, loquats, cherries, morello cherries, one of many among countless classes of berries and, also, of almond.

An attractive, ornamental fruit, the red coloring of the strawberry strongly excites desire and the pleasures of taste. Its suave, delicate, seductive bittersweet flavor and juicy, inviting-to-bite body make the strawberry unsurpassable as a fruit. There is no one who does not like them.

As Eurico Teixeira says, the strawberry is a veritable "dessert delicacy," whether simply washed, with sugar and with cream - whipped or not; with sugar and milk; sprinkled with lemon or orange juice or with liqueurs or rose water; dipped in port wine or mixed with sugar, or even in sweet vinegar.

Moreover, strawberries, especially the ones considered overripe for *in natura* consumption, may be made into jams and jellies, purées and syrups that, in addition to being eaten as desserts, may be used in the preparation of sweets and desserts: pies, cakes, puddings, mousses, charlottes, ice creams, meringues, etc.

The story of the strawberry may be thought of as a long one in the pursuit of perfecting flavor and pleasure. As Pio Corrêa tells us, although there are many native species of the strawberry in the majority of the temperate regions, the fruit at present cultivated comes from the crossing of and union of several species, among them some found only in their wild form in the Americas.

Even though there is, throughout Europe, a record of the various species of strawberries known and appreciated since olden times, it is presumed that they became part of family gardens in only about the 15th century. Nonetheless, despite their providing fruit of excellent quality and fabulous aroma, the strawberries of that time were small and their production quite irregular.

This wild strawberry continued delighting palates for some time, but also frustrated those who wanted to eat them in abundance – until the European colonizers arrived in North America. There they encountered a native strawberry that was extraordinarily productive and vigorous (*Fragaria virginiana*), which, early in the seventeenth century, was taken to the European continent.

A century later, a new and fortuitous find: farther south, in the lands that are now Chile, the colonizers stumbled upon another native species, a larger one (*Fragraria chiloensis*) that had long been cultivated by the natives. It, too, was taken to Europe.

Pio Corrêa believes that the casual crossing among these species of *Fragraria* in the European nurseries originated the matrixes for the countless varieties we have today. These improved strawberries returned to America to participate, in mid-19th century, in the United States, in the first studies and experiments in the search for the genetic improvement of the species of fruits known in America. The same author believed that this was not simply a "sensational improvement of strawberries," but, indeed, a true "result of a definitive effort in genetics."

Currently, in Brazil, as in almost the whole world, there are innumerable, different varieties of strawberries being cultivated and, in general, the ones that adapt best to the region are utilized. The state of São Paulo stands out as the major domestic strawberry producer; the southern edge of the state of Minas Gerais is also an important strawberry-producing center.

Among the coconut palms that grow and produce abundantly in Brazil and bear fruit that is generally much appreciated and of high industrial and commercial value, the outstanding one is that known commonly in the North of Brazil as the macaubeira and in the South as coqueiro-de-catarro. The synonymy of its folk names is vast: mucajá, mocujá, mocajá, macaúba, macaíba, macaiúva, bacaiúva, bocaiúva, umbocaiúva, imbocaiá, mbocaiúva, or mbocaíba."

FRUTAS DO BRASIL
Eurico Teixeira

MACAÚBA

Scientific Name: *Acrocomia aculeata* (Jacq.) Lood. ex Mart.
Botanical Family: Palmae

Origin: Brazil - Northern to Southeastern forests.
Plant Features: A palm, up to 15m in height, with an erect straight trunk covered with the remains of old leaves and with many dark spines on its surface. Curly, thorny leaves up to 1m in length. Small, yellow flowers in clusters up to 80cm long appear from October to January.
Fruit: Globular, smooth, yellowish-brown when ripe. Yellowish pulp with 1 oily seed. Fruition is from September to January.
Cultivation: Propagation is by seed. The plant requires a sandy substrate rich in organic matter, in a shady location.

Acrocomia aculeata (Jacq.) Lood. ex Mart.

Acrocomia intumescens Drud.

"Over millions of years, palm trees have adapted to the varied climates and soils. The majority prosper in the hot, humid equatorial climate. Others do well in arid, semi-desert climates and thus withstand prolonged dry seasons. Others again, leaving their tropical belt, withstand temperatures below zero. They grow in soil that is good for agriculture. They have also adapted to acid, siliceous, sterile soils, in which no other economically cultivated plant would flourish. Not only do they grow in swamp land, but also on dry rock land with no humus at all."

PALMEIRAS DO BRASIL
Gregório Bondar

The macaúba appears in vast regions of Brazil, extending from the states of São Paulo and Rio de Janeiro, going through Minas Gerais and all of the Midwest, Northeast, and North of Brazil, crossing borders to reach into Central America. According to Pio Correa's dictionary, some scholars believe the macaúba has even reached Mexico.

With this dispersion and the similarities among the palms of the *Acrocomia* genus, to which the macaúba, the macaíba, and the mucajá belong, it is only natural that there are different regional names to designate them and that they become confused.

According to Gregório Bondar, the *Acrocomia* genus embraces fifteen different species, ten of which are considered Brazilian. In every case, they are tall, robust palms, whose slender, elegant, erect trunks rise up to twenty meters in height and are usually covered by fierce thorns, especially near the top. In their natural habitat, they are noted for their exuberant beauty as they grow in clusters, often together with other palms such as the carnaúba.

Utilization of the fruit of the macaúba (the *Acrocomia aculeata*, which has the most thorns and is concentrated in the states of Minas Gerais and Goias) as well as of the macaíba (*Acrocomia intumescens*, a shorter tree with a sort of belly on its trunk, which is found mostly in the northeastern states) and of the mucajá (*Acrocomia sclerocarpa*, which has fewer thorns and is more frequent in the North) is also very similar.

In all three, the pulp enveloping the kernels of the coconuts which these palms produce is frequently eaten *in natura* by the local populations. However, they also yield good quantities of oil, utilized mainly by the soap industry, but which is also used in cooking, in home remedies, and as lamp oil.

According to Pio Corrêa, of greater importance is the transparent, colorless oil of the kernels, especially of the macaúba, which, in addition to its high yield, is light and edible, a good substitute for olive oil. Corrêa also calls attention to the fact that in the region of Diamantina, in Minas Gerais, the outer coat of the coconut – which is quite thick and hard, like bone or ivory – is used to make decorations and ornaments, such as rings, cuff links, chains, and so on.

Despite their abundant production of fruit and their other qualities, these native Brazilian palms of the *Acrocomia* genus, growing wild throughout most of the country, have been exploited in only the most rudimentary, cottage-industry-type fashion, far short of their economic potential.

Alternatives for their better utilization are being studied through experiments carried out by EMBRAPA's Agricultural and Livestock Research Center for the Savanna, in Brasília, where several tasty, nutritive recipes for sweets, meals, jellies, and candy have been developed, utilizing the pulp of the macaúba coconut.

Acrocomia sclerocarpa Mart.

MORGOTE TANGERINE

Scientific Name: *Citrus aurantium X reticulata* var. murcote
Botanical Family: Rutaceae

Origin: Asia and Australia.
Plant Features: A medium-sized tree up to 4m in height with spiny branches, shiny dark-green leaves and white flowers in small clusters.
Fruit: Rounded, flattened at the ends with a thin skin, its main characteristics being its firm adherence to the pulp. Fruition is from April to October.
Cultivation: The plant prefers hot climates, adapts to any type of deep, well aerated soil. It is propagated mainly by grafting. Planting should occur at the beginning of the rainy season.

The tangerines belong to the botanical family of the *Rutaceas*, which includes more than nine hundred different species, and to the *Citrus* genus, like all the fruits known as citrous. Like the other fruits of the *Citrus* genus, there are many tangerine, or mandarin orange varieties with its hybrids produced and sold in all the domestic and international markets.

Their origin is the same as that of the other citrous fruits: very different from those cultivated today, it is supposed that the first wild tangerine species tasted by man were found in Asia, probably in China.

Of the citrous fruits, the tangerine is, perhaps, the most perfumed and the most pleasant to eat *in natura.*

One of its main characteristics, other than its lovely bright orange color, is the ease of peeling this fruit. Its juicy segments are enveloped by a fine pellicle, and, depending on the species and variety, the segments can be separated easily.

Consequently, the act of eating the tangerine often necessarily becomes a rather informal affair, that is, without the use of anything other than one's hands. Like other citrus fruits, the porous tangerine skins contain an abundant, strongly perfumed, essential oil that has the same properties as the juice, however in much higher proportions. When handled, this unmistakable essence impregnates one's hands for a long time. It is impossible to disguise!

In Brazil, the commonly cultivated tangerine varieties and, therefore, the best-known are the Cravo, Poncã, Mexirica do Rio, and the hybrid, the Murcote.

The last, called the morgote, tangerina-morgote, or laranja morgote in Brazil, is a variety obtained by crossing the orange (*Citrus aurantium*) with the tangerine (*Citrus reticulata*), which is said to have appeared on the North American market in the late fifties. The morgote's principal characteristics are that it is one of the smallest tangerines cultivated commercially, is sweet and very juicy, has a thin skin, and adheres to the segments more than other tangerines.

MURUMURU

Scientific Name: *Astrocaryum murumuru* Mart.
Botanical Family: Palmae

Origin: Brazil - Amazon region, mainly Pará State and Marajó Island.
Plant Features: A palm that grows in clumps, with a trunk that may reach 10m in height. The trunk is covered in black thorns and characteristically marked by the bases of old leaves attached to the surface. Fronds are silver-green, up to 4m in length. Very small yellow flowers.
Fruit: Pear-shaped, with short fine thorns, red when ripe. Yellow pulp. The stone encloses a hard kernel. Fruition is from February to April.
Cultivation: Grows spontaneously in the forests of Pará in damp areas. Also common on the Island of Marajó.

The murumuru palm is common in the Amazon, especially in the state of Pará in swamps or very damp soils; it can also often be found in the low grassy lands of the island of Marajó.

Not a very tall palm, its main characteristic is a trunk covered with sharp, black thorns, like the brejaúva, another palm of the *Astrocaryum* genus. Its thorns, practically horizontal to its trunk, can grow to twelve centimeters in length, giving the innocent murumuru a rather aggressive appearance.

Another characteristic particular to this palm is the fact that its laden clusters of nuts grow upward.

The fruit of the murumuru is a coconut that is reddish-colored when ripe. The nut contains a hard kernel enveloped in an edible, slightly sweet, yellowish pulp. For Pio Corrêa, its smell and taste resembles that of the melon.

The quite oily kernel of the murumuru is also edible and, after processing, is used as a raw material for margarines, its main commercial use.

MURICI

Scientific Name: *Byrsonima basiloba* A. Juss. " crassifolia (L.) Kunth ???

Botanical Family: Malpighiaceae

Origin: Brazilian scrub land regions.

Plant Features: A tree, up to 10m in height with a cylindrical trunk, dark-colored, rough bark and a narrow crown. Leaves are green, rigid, with white pilosity on the under side. The orange-colored flowers appear from January to March.

Fruit: Round, orange-colored, with a meaty translucent pulp and cream-colored seeds. Fruition is from April to June.

Cultivation: The seeds germinate in a clayey substrate, requiring a shady location. Growth is slow. The species is also used as an ornamental.

The muricis of Brazil are many and varied; most are plants of the botanical family of the *Malpighiaceae*, to which the highly regarded acerola, or Barbados cherry, also belongs. Native to Brazil, these plants may be found throughout the continent and bear the popular names given them in the various regions.

The different varieties of the murici are distinctive for their colors and the places where they occur. Therefore, some are known as the yellow murici, white murici, red murici, white flower murici, red flower murici, the plains murici, the meadow murici, the swamp murici, and the beach murici, among many others.

Most of the muricis grow spontaneously in practically all of the Amazon, where large bush or tree-like species may be encountered in abundance. During fruit-bearing time, the green woodlands are speckled by the yellow fruit.

But the muricis are not exclusive to the forest;

some dot the hillsides of the Southeast, the scrub lands of Mato Grosso. and the north and northeastern coasts of the country.

Among the regional Brazilian fruits posted on the early-morning boards of street vendors – especially in the state capitals of the Northeast – the murici certainly holds a special place and, depending on the time of year, stands on its own together with such important standards as the graviola, jambo, caju (cashew), and pitomba.

Here, two murici species are focused on. One, the most widely dispersed, the best known, and the provider of the most appreciated fruit is the *Byrsonima crassifolia*. It is a small tree, a maximum of five meters high, whose fruit is extremely tasty and delicately perfumed.

Cited by Pimentel Gomes, Renato Braga made mention of the many uses regional people learned to make of the murici, many of which come from olden times. According to Braga, along the northeast coast one of the most common uses of this fruit is *"a very fatty, highly nutritive mash known as* ***cambica de murici****."* *"The fruit,"* says Braga, *"is mashed in water; it loosens easily from its pulp, which, when dissolved, mixed with manioc meal, sweetened or not, constitutes one of the most important food sources for the poor beach vendors."* Moreover, this mash, more diluted and not mixed with the meal, becomes a refreshing drink and is an ingredient for excellent ice creams and sweets.

Everything leads one to believe that it was the murici that was discovered in Bahia by the itinerant priest, Gabriel Soares de Sousa, at the end of the 16th century. In his description, he said that the murici was a small, very dry tree that, springing from weak soils, produced soft, yellow fruit, smaller than cherries, edible, and of a flavor and smell similar to the *"Alentejo cheese."*

The species known as the murici-do-campo (meadow or field murici, the *Byrsonima basiloba*) differs from the first due basically due to its size. A tree of average size and twisted limbs, typical of the scrub lands, it reaches ten meters in height. Its leaves are also larger than those of the *Byrsonima crassifolia's* and are covered with tiny hairs, giving it a velvety feel. Its fruit and uses, however, are much like those of the common murici.

Byrsonima basiloba A. Juss.

SOUARI NUT

Scientific Name: *Caryocar brasiliense* Camb.
Botanical Family: Caryocaceae

Origin: Brazil - Scrub land regions.

Plant Features: A tree, up to 10m in height with a twisted trunk and a rough, grooved bark. The hairy leaves are formed of 3 leaflets with serrated edges. Large yellow flowers appear from September to December.

Fruit: The bright yellow pulp encloses a hard stone, formed of a large number of small spines. Fruition is from January to April.

Cultivation: Planting from seeds occurs during the rainy season. The plant prefers hot climates, ideally the Northern, Northeastern and Mid-Western regions of Brazil. Saplings grow slowly. Each plant provides an average of 6 thousand fruits per year.

In December, the air in the region as well as around scrub-land kitchens is redolent with the perfume released by a little fruit called the Souari Nut, piki or pecka nut.

The first to arrive are the wild animals, bees and other insects, birds of all sizes, the various-sized rodents and mammals of the scrub land; peccary, agouti, armadillo, cavi, deer, macaws...

Then come the men; whole families arrive to start harvesting the nuts, which drop easily from the stems of the native trees, to be found throughout the scrub lands and forests of central Brazil. Soon after, the fruit may be found everywhere, in small villages or in the main streets of large cities such as Goiânia (Goiás), Brasília (Fed. District) and Belo Horizonte (Minas Gerais), where peddlers sell the recently gathered nuts.

The fruit, the size of an orange, is ripe when its skin, which remains always the same greenish-yellow color,

softens. Cutting it open, from one to four stones are to be found, tender spiny kernels surrounded by a yellow, white or pink pulp, the plant's real attraction. The only hindrances are the tiny, fine, sharp spines existing in the center of the stones, and much care must be taken when chewing on them to eat the pulp.

The Souari is highly appreciated in the regions where it grows; rice, chicken and beans with Souari are favorite dishes in regional culinary art. Souari liqueur is nationally famous and there are a variety of recipes for sweets aromatized with the flavor of Souari. There is, however, a great deal of controversy, as some people cannot so much as smell the penetrating odor of the fruit, while others really love its flavor and, when it is in season, will consume vast quantities, satisfying their craving built up during the rest of the year. Currently, it is possible to find Souari pulp or even the entire fruit frozen, but the true connoisseur claims that there is nothing like the Souari picked and eaten when it is ripe.

With a high calorie count, besides the perfumed, unique flavor that leads it be used as an ingredient and flavoring for preparing a number of dishes, the Souari pulp contains a large volume of edible oil (roughly 66%) and is rich in vitamin A and proteins (13.5%). Therefore, it also becomes an important element in complementing the diet and nutrition of the population.

The Souari nut, due to its high oil content and its chemical characteristics, may also be used to advantage by the cosmetics industry for manufacturing toilet soaps and creams. Unfortunately for the plant itself, its wood also produces an excellent charcoal, which has been widely exploited. Over recent

years the fires of lime kilns and land clearing have caused a considerable reduction in the number of native trees in the scrub lands, and the tree that bears a such well liked and nutritious fruit is already on the brink of extinction.

In an attempt to save the Souari and its nut, technicians in EMBRAPA's Federal District Scrubland Farming Research Center have, together with other plants native to the region, been studying and evaluating the possibilities of producing seedlings, which are already being distributed. The objective is to initiate a campaign for the rational exploitation of the Souari, using it especially in reforestation areas.

In addition to the scientists concerned with the preservation of the species and its native lovers, the perfume of the Souari has always awakened passions. Patativa do Assaré, a popular poet from Ceará was one of them. Talking about the forbidden fruit, he was reminded of the Souari: *This fruit of sin / Seems to have one thing about it / That people seeing it, get / A desire to eat it. / Mister, I'm not sure / But I have an idea / What other fruit / Pleased our senses / As this forest fruit / That folk call piki.*

The fruit has this virtue: the flavors and the senses they awaken make one remember and revive old stories.

The Souari, as it is best known in English and as it has been referred to this far, is identified in a dictionary by Pio Corrêa as the *piquiá-bravo* (wild piki). Whatever its name, it is the fruit that grows from an average-sized tree and is native to the Brazilian scrub lands. In the same book by Pio Corrêa another fruit is mentioned, very similar to the former, which is named *piquiá-verdadeiro* (true piki). And this one we are simply going to call *Piki*.

Caryocar villosum (Aubl.) Pers.

If the Souari flowers and bears fruit in the scrub lands, the Piki is typical of the Amazonian forest, on dry land. Like its cousin in the scrub land, the Piki is very important to the hinterland populations who still preserve the habit of cultivating it. However, it is rarely cultivated in the larger Amazonian cities, although some trees are encountered ornamenting the roads and city squares of Manaus (Amazonas).

Both of the fruits, Souari and Piki, have the same characteristics: the pulp of the Piki fruit, cooked or raw is also edible, and comprises a source of fat and food for the hinterland populations. An infusion of the bark of the tree is believed to have anti-thermal and diuretic properties.

The main difference between the two species lies in the overall dimensions of the plants. As opposed to the tree that bears the Souari, the Piki tree is very tall, reaching as much as 40 meters in the dense forest and has a rounded base with a diameter that may reach five meters. Therein lies one of the secrets of nature: the capacity species have for adapting to their environments.

SURINAM CHERRY

Scientific Name: *Eugenia uniflora* L.
Botanical Family: Myrtaceae

Origin: Forests from the states of Minas Gerais to Rio Grande do Sul.
Plant Features: A tree, may reach 10m in height with an irregular trunk, well branched, with a reddish color and a bark that may peel occasionally. The oval leaves are reddish-colored when young, eventually turning a bright green, shiny, with a characteristic odor when crushed. The aromatic white flowers appear from August to November.

Fruit: Round, flattened top and bottom with longitudinal grooves, colored orange and bright red when ripe. The pulp is red and meaty, enclosing a greenish colored seed. Fruition is from October to January.
Cultivation: Thrives in regions with a hot, damp climate. No requirements regarding sunlight. Propagation may be from the seed or by grafting. No large scale planting is done in Brazil.

Eugenia uniflora L.

The word *pitanga* comes from the Tupi, meaning deep red. This Brazilian cherry is, indeed, a deep cherry red, almost purple, at times almost black, and extremely tasty, refreshing, cooling and – as was said long ago, "a great balsam for the blood."

The sweet taste of the pitanga's slightly acid, characteristically perfumed pulp holds a specific place in Brazilian taste. And picking it from the tree is part of Brazilian culture and sentiment. Its tiny delicateness, its segmented shape and redness are symbols of the Earth.

Originating in Brazil, the pitanga is to be found everywhere by anyone waiting to taste. It stretches from the Northeast to Rio Grande do Sul and crosses the border into some regions of Uruguay and Argentina.

Either small or large, the pitanga, when cultivated, is typical and suitable for backyards and homes or small farm orchards, where this very ornamental tree is used to form attractive fences and gardens.

Its flowering is abundant, white, and perfumed.

When it bears fruit, it becomes transformed, attracting attention from afar, because its branches become a bright red, attracting many birds, children, and adults that have forgotten they have grown up, all of them delighting in the flavor of the ripe fruit.

In addition to eating the fruit *in natura*, Brazilians have created many recipes for juices, refreshments, jellies, and sweets, besides the famous *"pitanga liqueur or cognac,"* to which aphrodisiac properties are attributed. The latter, also known as the "tropical cognac," immortalized in the book, *Açucar* (Sugar), written by the Pernambucan, Gilberto Freyre, is one of the regional drinks most characteristic of the Brazilian northeast, along with sugar cane juice, rum mixed with honey, and various juices and wines made from native fruits.

Owing to its size and ease of handling and cultivation, and its resistance to urbanization, the pitanga is suitable for planting in squares, parks, and along sidewalks, thus contributing to the beautifying of cities and the minimum sustenance of what remains of the avifauna.

Pitanga leaves are also highly perfumed and, together with mango and cinnamon leaves, are traditionally related to Catholic and syncretic cults and rituals and Afro-Brazilian practices of Bahian Candomblé.

Eugenia calycina Camb.

Often these leaves are used to cover the ground of the terreiros (Candomblé places of worship) and streets, in processions and on days of celebration.

There are also some native, regional varieties, such as the pitanga do cerrado, or savanna pitanga (*Eugenia calycina*). Found in the scrub lands, this pitanga is taller, more upright, and the fruit does not have the characteristic external furrows of the common pitanga.

Currently, outside its region of origin in South America, the pitanga tree is cultivated in the southern United States, in the Caribbean islands, and even in India and China.

Unfortunately, as Pimentel Gomes has already said, since the fruit is little cultivated on a commercial scale, it is not ordinarily found in the market places of south and southeastern Brazil since its distribution in large cities is sparse and irregular.

Nonetheless, some juice and ice cream companies in the Northeast already have pitanga orchards, whose fruit is used for the production of bottled pitanga juice and frozen pulp.

PERINHA

Scientific Name: *Eugenia lutescens* Camb.
Botanical Family: Myrtaceae

Origin: Brazilian scrub land regions.
Plant Features: A bush up to 1m in height, with rigid, oval leaves, showing a whitish pilosity when young. Aromatic white flowers.
Fruit: A round, rough greenish fruit when young, becoming smooth and yellow when ripe. A yellow, meaty, aromatic pulp surrounds 1 to 3 seeds. Fruition is from September to November.
Cultivation: Grows spontaneously in scrub lands, especially in the state of Minas Gerais.

Traveling through the hinterlands of Brazil, across the states of Minas Gerais, Goias, Mato Grosso, and Mato Grosso do Sul, means traversing vast areas of vegetation characterized by short, twisted trees, set in the midst of a carpet of herbs and herbaceous plants of various different aspects. This is the Cerrado (savanna).

However, the people of the Mid-West of Brazil know how to recognize the existence of and distinguish between the various forms and faces of this

irregular scenery. Indeed, these are the savannas, the scrub land, barren land and meadow lands that extend for thousands of kilometers.

It is also a landscape that changes greatly in summer and winter. During the dry season, the so-called winter, the pasture turns yellow and dry and most of the trees and bushes lose their leaves. Then, the first summer rains come and everything turns green.

In spite of its rustic appearance, landscape dissimilarities and the drastic changes brought by climate and seasons, the *Cerrado* (savanna) yields a profusion of flowers, fruit, and color all year long. As Professor Leopoldo M. Coutinho says, *"despite the apparent monotony of vegetation, the flora of the Cerrado is enormously rich,"* losing – in the number of species – only to the wet tropical forests.

And there, in the midst of the savanna, the perinha (little pear) is born and grows. A low bush of the extensive *Mirtaceas* family, so typical of Brazil, the perinha is a fruit much resembling the uvaia (uvalha eugenia), only a bit larger. Its yellow, plump, and fragrant pulp is slightly sour and may be eaten *in natura* or as a refreshing juice or jellies, as in fact almost all the delicious and juice-filled fruits of this family, may be.

PINDAÍBA

Scientific Name: *Duguetia lanceolata* St. Hill
Botanical Family: Annonaceae

Origin: Midwestern, Southeastern and Southern regions os Brazil.
Plant Features: The tree can reach 20 m in height, rough bark, brown to grayish colored. Shiny yellowish-green leaves. Flowers are red when in bud, opening to pale pink. Flowering from October to November.
Fruit: Globular, formed of a dense group of brown seeds surrounded by pulp, resembling scales with a meaty consistency and wine-colored when ripe. Fruition is from March to May.
Cultivation: The plant occurs naturally in high, forested regions, always in well drained soil. Propagation is by seeds. Slow growth.

The pindaíba is a fruit belonging to the *Anonaceas* family and is, therefore, related to the araticun or custard apple, the pinha or sweetsop, the biribá, the graviola or soursop, and the pimenta-de macaco.

In Brazil, many different plants of this family are known by the name pindaíba. In addition to the pindaíba in the photo above, whose scientific name is *Duguetia lanceolata*, several other plants of the same family are called pindaíbas.

However, the fruit of the pindaíba-vermelha, the

pindaíba-reta, the pindaíba-do-brejo, and the pindaíba d'água (the red, straight, swamp, and water pindaíbas), for example, are not shaped like the "pinhas," reminding one more of the shape of the pimenta-de-macaco fruit. The pindaíba in the photo, on the contrary, has a shape and size similar to the sweetsop or fruta-de-conde, the *Ananona squamosa*. Despite the similarity, they are not mistaken for one another. Moreover, anyone who has seen the tall, slender pindaíba knows it is a distinctly different species.

The pindaíba has a rustic, very pretty, special look. As it ripens, its green coloring turns to tones of red until it becomes a violet blood-red.

In the interior of the state of São Paulo, it is said that children actually drool, while eagerly waiting for their parents and adult friends and relatives to return from the woods, where they gather the fruit.

This is told by people who have sucked the pinkish-red pulp covering the seeds. They also say that the pindaíba is more flavorful than the common sweetsop, although the layer of pulp is thinner.

Presumably, the expression *"estar na pindaíba"* is perhaps associated with the fact that the layer of pulp is very thin and has little substance: when a person says he "is on the pindaíba," it means he is broke and has no alternative but to eat pindaíbas, even though he knows they are not filling.

Native to and common in the mid-west, south and southeast of Brazil, the pindaíba grows mainly in the higher hill areas and in the Atlantic rain forest, like its homonymous relatives. Less found in these forests today, the pindaíba is a beautiful delicacy that is disappearing.

PIMENTA-DE-MACACO

Scientific Name: *Xylopia aromatica* (Lam.) Mart.
Botanical Family: Annonaceae

Origin: Brazilian scrub land regions.
Plant Features: A tree, up to 6m in height with a dark bark and a scant crown. The bright green leaves are hairy on both sides. White flowers, appearing between September and November.
Fruit: An irregular shape, greenish in color, with a red pulp containing dark, almost black seeds surrounded by a white aril. Fruition is from April to July.
Cultivation: The plant prefers sandy soils rich in organic matter. It is slow growing and propagates from seeds.

Also known as pimenta-de-negro, the pimenta-de-macaco (black or monkey pepper) is a member of the *Anonaceas* family and is, therefore, related to the araticun (custard apple), the pinha (sweetsop), the biribá, the graviola, and the pindaíba.

It grows spontaneously, basically in the savannas in central Brazil. Despite being widely found throughout the states of Minas Gerais, Goiás, South Mato Grosso, and part of São Paulo, the distribution of the pimenta-de-macaco is irregular.

The tree itself is small, not over six meters tall and, in Harri Lorenzi's opinion, is quite ornamental due

to its airy crown and long, narrow branches.

Its fruit resembles that of many other Xylopias, or Brazilian pindaíbas, that is, parts of the fruit are separate, making them seem like clusters.

When ripe, the pimenta-de-macaco fruits open into two parts, exposing a fine red pulp guarding its bright black, wrinkled seeds. This pulp, called aril, is greatly sought after and eaten by various species of birds.

Perhaps it is the aroma of this fruit that so attracts birds, not just its reddish-pink dotting the crown.

This aril that envelopes the seed and lines the fruit, like the seeds, has a slightly peppery flavor. In his dictionary, Pio Corrêa says that the pimenta-de-macaco seeds are aromatic and may be used as a seasoning, as a substitute for black pepper, in either their seed or ground form. Consequently, it is used as such in regional Brazilian cooking, which may be a reason for the plant's names.

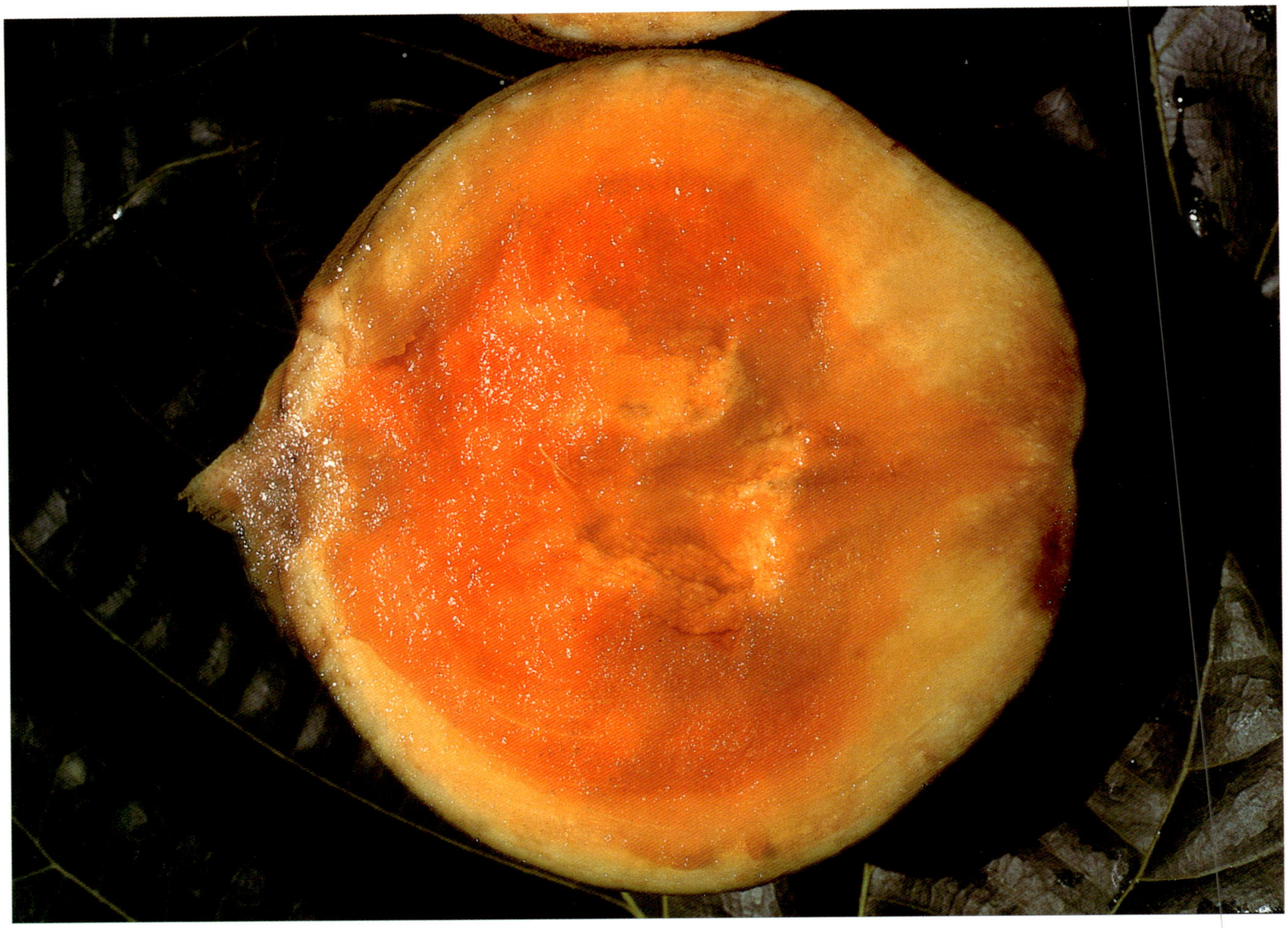

SAPOTA-DO-SOLIMÕES

Scientific Name: *Matisia cordata* H.B.K.
Botanical Family: Bombacaceae

Origin: Brazil - Amazon region as far as Colombia and Peru.
Plant Features: A large tree, may reach 45m in height. Large leaves up to 50 cm in length. The pinkish-white flowers appear from August to November.
Fruit: Oval in shape, characteristically covered by a "hood", a persistent portion of the flower. The orange-yellow pulp encloses 2 to 3 thick, hard greenish-brown seeds.
Cultivation: Grows spontaneously throughout the Amazon region, where it is widely cultivated.

Among the various plants popularly called sapotas, the sapota-do-solimões (Solimões sapote) is the one that is most different. More than the others, it is greatly appreciated for its fruit which, in addition to being edible, is distinguished by its round or oval shape and large size.

The outside of the sapota-do-solimões fruit is a thick, greenish-brown skin, while inside it has lots of succulent pulp replete with fine orange fibers. When in season, this fruit may commonly be found in the street markets of some Amazon towns, since its delicate, sweet, flavorful pulp is much liked and sought after by the locals.

A large tree, the sapota-do-solimões grows to a height of 40 or 45 meters in the forest, where its heavy dark green crown of round leaves provides generous shade.

It is a spontaneous species, native to Brazil's western frontiers with Peru and Bolivia, and prefers dry ground forests that are not flooded. It is also cultivated in the upper Solimões region, which is probably the reason for the name by which it is best known.

Although the sapota bears heavily, its cultivation is limited to the areas to which it is native. Studies and experiments carried out by the National Research Institute of the Amazon (INPA) researchers are considering the possibility of including the sapota-do-solimões among the Amazon fruits to be cultivated on a large scale.

MONKEY POT

Scientific Name: *Lecythis pisonis* Camb.
Botanical Family: Lecythydaceae

Origin: Brazil - Atlantic rain forest.
Plant Features: A tree, up to 30m in height, with a trunk that may reach 1m in diameter and a dense crown. Leaves are characteristically pink when young, later turning green. Large, purplish lilac-colored flowers appear abundantly from September to October.
Fruit: Rounded, with a brown colored tough, thick rind. When ripe, a characteristic "lid" in the bottom portion opens, releasing the tasty edible seeds (nuts). Fruition is from August to September.
Cultivation: Propagates from the seed. Should the fruit have been picked before it is ripe, leave it in the sun to open and release the seeds. Requires a clayey soil, rich in organic matter and a shady location.

A large number of trees are known by the name of sapucaia in Brazil. They belong to the *Leciti-daceas* family, that of the majestic Brazil nut, or castanheira-do-pará.

Most of the sapucaias are characterized by the peculiarity of their fruit. Shaped like urns and having hard, woody shells, these pyxidia contain a good number of edible, much enjoyed nuts. When the fruit is ripe, the urn lids or opercula open, dropping their contents all over the forest.

Native to Brazil, the sapucaias and their fruit were already well-known and utilized by the indigenous peoples, when the Europeans first arrived in the 16th century. The colonizers were attracted to the tree's high points – useful, exotic, ornamental – and, impressed by these peculiarities, wrote interesting descriptions detailing its configuration.

As Eurico Teixeira tells us, traveler Péro de Magalhães Gândavo, described this fruit as a large, very hard coconut, full of sweet, extremely flavorful nuts. To him, the fruit did not seem to be created by nature but, indeed, by some "artifice of human industry", since their downturned mouths are covered by "capadoiras" (opercula) that fall off by themselves, allowing the nuts also to fall and scatter.

The aromatic, oil-rich, nutritive sapucaia nut may be eaten raw, boiled or roasted. It may be used to substitute European walnuts, almonds, and chestnuts, for it lends itself well to use in desserts, sweets, and main dishes.

The tree is characteristic of the Atlantic rain forest and, therefore, can be found from Ceará to Rio de Janeiro, but particularly in southern Bahia and northern Espírito Santo. It also occurs in the Amazon. In some cases, deep in the forest, this tree displays the greatness nature bestowed upon it as it reaches up to more than thirty meters.

Sapucaia blossom time is party time for the bees. The sapucaia is transformed: its green crown is covered by a pink cape, a wondrous sight given us by its thickly set flowers. Highly perfumed, the blossoms, tending toward purplish tones, take over the crown of the tree and its new leaves, which also sprout colored lilac or rose.

Slowly the leaves turn green and the fruit begins to form. When empty, the receptacles for the nuts are made, by man, into useful objects and ornaments: gourd-like bottles, small buckets, vases, pots, plates, lunch pails, and so on.

Wild animals love the sapucaia nuts; monkeys especially do and go after the nuts up high in the trees before they fall out of their "pots." Indeed, a curious fact involves these animals and this fruit, which is also called the "monkey pot" or "monkey lunch pail."

As Pio Corrêa and several other scholars tell us that, when the young, inexperienced monkey sees a sapucaia open and full of nuts, it goes greedily to the pot and sticks his hand into it to grab all the nuts at once. Then, when it tries to withdraw its hand, it either can't or hurts himself, because its hand, full of the nuts, will not come back through the narrow opening of the sapucaia. The old monkey does not make this mistake. With the wisdom gained from having made the same mistake, it uses the tips of its fingers to lift the nuts out one by one, while it is chewing others.

So it seems that the monkeys' penchant for the fruit of the sapucaia gave birth to the proverb, "Macaco velho não põe mão em cumbuca!" (or: A wise old monkey doesn't stick its hand into a pot).

"I am not going to sing a hymn to the worth of the sapucaia, but ... its entire fruit may be utilized: its bucket, or pot, is a lunch pail, a pot, a vessel, whatever; the pulp is edible, medicinal, highly nutritive; the seeds or castanha-de-sapucaia (the nut) have a wonderful taste, produce magnificent oil, are a substantial food; the leaves, when turned to ash, may be used as fertilizer; the roots have therapeutic use."

FRUTAS DO BRASIL
Eurico Teixeira

"A species known as kumã-uaçu or sorva grande, it produces abundant sap, rich in rubber, and its fruit is very tasty and sweet. The epicarp of the fruit is tough, so that the pulp containing the edible part, the seeds, must be pulled out off it."
Barbosa Rodrigues (1891) quoted by Paulo Cavalcante

SORVA

Scientific Name: *Couma utilis* Muell. Arg.
Botanical Family: Apocynaceae

Origin: Brazil - Amazon region.
Plant Features: A tree, with a height varying from 6 to 20m and a wide dense crown. The plant holds an abundant supply of a milky, potable latex. The brilliant green, rigid leaves grow up to 10cm in length. Small pink flowers, in clusters of 15 to 20 blossoms, appear from May to August.
Fruit: The globular, green fruit, turns dark brown when ripe, and has a thin skin, containing a viscous, milky juice. The pulp is mucilaginous and yellowish in color.
Cultivation: Occurs spontaneously in the Amazon region.

The sorveiras or Brazil sorva are diverse and very common in the entire Amazon region, especially in the states of Amazonas, Pará, Amapa, and Rondonia, and stretch into the Guyanas, Colombia, and Peru.

Wild sorvas are found in the midst of thick virgin forests, and on marshy or dry land. Some varieties are spontaneous in fields or grasslands and in secondary forests; they are frequently grown around Manaus.

The fruits of all the sorva varieties are the size of lemons and are at first green, ripening to a dark brownish color. Besides their good, sweet flavor and being a nutritive food for the regional peoples when consumed *in natura* or as a beverage, the fruit of the sorva tree is not the only product that can be extracted from it.

From the trunk, especially of the *Couma macrocarpa* species (sorva grande) and the *Couma utilis* (sorva pequena), good quantities of a thick, white, viscous, edible, sweetish latex may be extracted. This latex may be drunk pure, but diluted in water. In this form, it is often taken in place of cow's milk, to be drunk with coffee, and serves as an ingredient for porridge.

In the forest, for example, it is common to see rubber tappers going out to work without taking any food. That is because from trees like the sorva and their nutritive latex, the inhabitant of the land encounters his daily sustenance.

Tapped from the trees by a process similar to that used for the rubber tree, the sorva latex is also widely used as a chewing gum ingredient. After extraction, the latex solidifies and is sold in large compact blocks, usually for export. According to Paulo Cavalcante, the exploitation of sorva for this purpose was intense in the forest, but has dropped sharply in recent decades.

Sorva latex is also used industrially to produce gums/glues and varnishes. Moreover, for many years the natives of the Amazon have known that, in addition to its food uses, sorva latex has insulation properties, as it is resistant to weather and humidity. Coagulated and mixed with other substances, this latex is also much used as a caulking material for vessels and for whitewashing houses in the Amazon.

GRAPE

Scientific Name: *Vitis vinifera* L.

Botanical Family: Vitaceae

Origin: Europe and the Middle East.

Plant Features: A vine with a thick, though stem, green when young, darkening with age. Large leaves divided into five lobes with fine whitish surface hairs. Small, green-cream colored flowers.

Fruit: Globular or elipsoid, the fruit can be white, green, yellow, pink, red or bluish, depending on the variety. The watery pulp can contain as many as 4 dark seeds. Frution is from October to February.

Cultivation: The plant prefers dry climates with a temperature of around 22°C, adapts to various soil types, except for damp, poorly drained ones, which must be rich in organic matter. The soil has a strong influence on the quality of both grape and resulting wine. Propagation is by means of layering or grafting.

The grapevine has been one of the best-known fruit bearing plants since ancient times; it is even found in fossils dating back to before man's arrival on Earth. Nonetheless, cultivation of the grape, a continuous and long affair – some vineyards have a 150-year history – became possible only in communities that had abandoned nomadism.

Giving up his nomadic ways, man little by little set about perfecting the agricultural techniques for growing and processing the grape. It may even be said that the history of the grape is a summary of the history of western man.

Initially, the cultivation of vineyards was only for family purposes and was developed for commercial purposes and for export only, when navigation developed in the Mediterranean.

Once again returning to Pio Corrêa's accounts, around the year 600 before Christ, man learned to prune the grape vine to obtain abundant, delicious fruit, an absolute leap toward improving production techniques. Legend also says that it was the donkey, not man, that first discovered the power of pruning, when it ate the tendrils and green leaves of a vine.

The belief is that grapes originated in Asia and were introduced into the Italian peninsula and Europe by the Greeks. In turn, it was the Romans who transformed grape growing into a lucrative business and filled the Mediterranean landscape with vineyards. Grapes, at that time and for centuries after were used mainly for making wine.

Since the grape was a difficult fruit to transport and preserve and demanding of care, man soon learned to make use of its valuable beverage qualities. Due to the fruit's high concentration of sugar,

"In the Gospels, the vine is a symbol of knowledge, wine having been chosen to symbolize the blood of the Creator when celebrating Mass. The vine symbolizes human life and wine is considered to be a divine gift. Its birthplace is the Orient, from where it was brought to Europe. During the time of the flood, Noah already planted vines and drank too much wine, according to the Bible. Among the Greeks, Saturn was regarded as a forerunner of viniculture in Crete; in Egypt, Osiris; in Italy, Bacchus."

DICIONÁRIO DAS PLANTAS ÚTEIS DO BRASIL
e das exóticas cultivadas
M. Pio Corrêa (1926)

its fermentation process generates an exceptional drink with variable alcohol percentages. It has become the best-known, most-consumed and most important of the fruit wines in existence. Grape wine is and has long been a product carrying with it enormous social and economic significance for western man.

Up to now, the greater part of the grapes produced are industrialized: it is calculated that about 80% of the total amount produced per year is made into wines or into other types of alcoholic beverages, such as brandy, cognac, armagnac, sherry, etc.; nearly 5% of the grape production goes to juices; another 5% becomes raisins; and 10% is eaten in natura, as dessert or for pure pleasure.

There are about ten thousand varieties of grapes, adapted to different types of soil and climate, which makes their cultivation possible in every region of the world. Being sensitive to the climatic and soil conditions in which they develop, grapes vary according to these conditions, presenting characteristics that distinguish the flavor, acidity, sweetness, shape, coloring and resistance of their skins, the size and number of their seeds, the shape and form of their clusters.

Regarding utilization, grapes are basically divided into two broad groups: table grapes and those for wines and other industrial purposes.

Grapes arrived in Brazil together with the first Portuguese colonizers. For a long time, according to Pio Corrêa, "the idea was that the environmental conditions here would never permit vineyard cultivation, the cultivation of an extremely delicate plant that could only thrive in some countries of Europe." Today, however, viniculture is now a great source of wealth for the country.

Ever since the beginning of the century, vineyards started to be installed in the southern and southeastern parts of the country by German and Italian immigrants. Many towns and villages formed, many grapes planted, definitively changing the cold hillsides of Rio Grande do Sul. It was in the townships of Flores da Cunha, Bento Gonçalves, Garibaldi, Santana do Livramento and Caxias do Sul, that the outstanding wine producers arose. Now, in addition to Rio Grande do Sul, the states of Santa Catarina, Paraná, and São Paulo (especially in the townships of São Miguel Arcanjo, Vinhedo, and Porto Feliz) also produce good grapes.

Nevertheless, since the early eighties, Brazil has undergone great changes in relation to grape production. Large domestic and international groups, through government incentives, began to invest in the development and installation of modern irrigation projects throughout the fertile, hot, dry valley of the São Francisco river. Innumerable enterprises producing juices, wines, and sweets were attracted to the region of the valley, increasing the offer of employment and greatly changing the socio-economic conditions of an entire population accustomed, for centuries, to the horrors of drought and scrub.

In a few years, the land of the licuri and mandacaru began the intensive production of fruits of all kinds: pineapples, papayas, avocados, figs, guavas, passion flower fruit, melons, watermelons, carambolas, mangoes, lemons, oranges, and, of course, grapes.

The agricultural and winegrowing potential of this region has long been estimated, by specialists, as, in the thirties, Pio Corrêa himself emphasized in his dictionary. The altitude, which varies between 300 and 500 meters above sea level, the superb arable lands, the always high annual average temperatures and the absence of rain and frosts are perfect conditions for the production of healthy, sweet fruit.

Passing through five northeastern and southeastern states of Brazil (Pernambuco, Sergipe, Alagoas, Bahia, and Minas Gerais), the São Francisco Valley vineyards today are the largest orchard and the largest fruit exporter in the country, capable of supplying a large part of the burgeoning domestic market.

We wish to express our special thanks to the following people and institutions:

Alice Giraldi, Antonio Fonseca, Antonio Marcos de Queiroz, Aracelis Stamatis, Aziz Nacib Ab'Saber, Carlo Cirenza, Carlos Hans Müller, Célia de Assis, Cibele Boni de Toledo, Dijalma Barbosa de Silva, Elizabete Lopes, Fábio de Brito Ávila e equipe da Empresa das Artes, Fernando B. de Almeida, Fernando Picarelli Martins, Flávio Carvalho Ferraz, Helena Maria Carvalho Ferraz, Heloisa Cavalcanti de Albuquerque, Ilza Maria Monico, João Caldas, João Ernesto Dierberger, Joaquim Teófilo Sobrinho, José Antonio da Silva, José Carlos Ribeiro, José Edmar Urano de Carvalho, José Fonseca da Silva, José Luiz Hernandes, Júlio Seabra Inglês de Souza, Laura Molina Meletti, Laura Tassara Ferraz, Luís Benedito Bacher, Luiz Carlos Silveira, Manoel Campos Filho, Manoel Carlos Chaparro, Marcos Eduardo Albertin, Mariana Pabst Martins, Mário Jardim, Maureen Bisilliat, Mércia Cardoso Batista, Murilo Carvalho, Nina Horta, Paulo Brandão Cavalcanti, Paulo Zanettini, Rejane Campos Araújo, Rubens Matuck, Samuel de Melo, Sérgio Romaniuc Neto, Sílvia Corrêa Chiea, Torquato Gaudêncio do Rego, Yvonne Sarué.

ABPM/Associação Brasileira de Produtores de Maçã (Fraiburgo/RS)

COOPAMA/Cooperativa Agrícola Mista Amazônica Ltda. (Castanhal/PA)

EMBRAPA/Empresa Brasileira de Pesquisa Agropecuária:
Centro de Pesquisa Agropecuária do Cerrado (CPAC - Brasília/DF);
Centro de Pesquisa Agroflorestal da Amazônia Oriental (CPATU - Belém/PA);
Centro Nacional de Pesquisa de Uva e Vinho (CNPUV - Bento Gonçalves/RS).

ESALQ/Escola Superior de Agricultura Luiz de Queiroz da Universidade de São Paulo (Piracicaba/SP)

Escola de Nutrição da UFBa/Universidade Federal da Bahia (Salvador/BA)

Fazenda Sete Lagoas Agrícola S/A (Mogi-Guaçu/SP)

FRUNORTE/Frutas do Nordeste (Vale do Açu/RN)

IAC/Instituto Agronômico de Campinas: Centro de Citricultura Sylvio Moreira (Cordeirópolis/SP)
Estação Experimental de Jundiaí/SP; Estação Experimental de Pariquera-açu (Vale do Ribeira/SP)
Fazenda Santa Elisa (Campinas/SP)

IBRAF/Instituto Brasileiro de Frutas (São Paulo/SP)

IEA/Instituto de Estudo Avançados da Universidade de São Paulo/SP

INPA/Instituto Nacional de Pesquisas da Amazônia (Manaus/AM)

Instituto de Botânica de São Paulo - Secretaria do Meio Ambiente do Governo de São Paulo/SP

IPA-UEP/Instituto de Pesquisa Agrícola (Goiana/PE)

Museu Paraense Emílio Goeldi (Belém/PA)

UFRPE/Universidade Federal Rural de Pernambuco (Recife/PE)

and to all those who, throughout Brazil, contributed to the well-being of this publication.

BIBLIOGRAPHY

ALMEIDA, S.P. de *et alli*. *Aproveitamento alimentar de espécies nativas dos Cerrados: araticum, baru, cagaita e jatobá*. Planaltina: EMBRAPA/Centro de Pesquisa Agropecuária do Cerrado, 1987.

——— e SILVA, J.A. da. *Piqui e buriti - importância alimentar para a população dos Cerrados*. Planaltina: EMBRAPA/Centro de Pesquisa Agropecuária do Cerrado, 1994.

ANDERSEN, O. e ANDERSEN, V. U. *As frutas silvestres brasileiras*. Rio de Janeiro: Globo, 1988.

ANDERSON, A. e MAY, P. "A palmeira de muitas vidas." *Ciência Hoje*. São Paulo, setembro/outubro de 1985. p. 58-64.

ANDRADE, C. D. de. *Menino antigo (Boitempo II)*. Rio de Janeiro: José Olympio, 1974.

ARAÚJO, P.S.R. de e MINAMI, K. *Acerola*. Campinas: Fundação Cargill, 1994.

BASTRYA, J. e CANNING, J. *A Gourmet's Guide to fruit*. Los Angeles: HP Books, 1990.

BAYMA, C. *Carnáuba*. Rio de Janeiro: Ministério da Agricultura/Serviço de Informação Agrícola, 1958.

BERNARDES, C. "Cascavel no olho do buritizeiro." *Globo Rural*. São Paulo, julho de1989. p. 114.

BISILLIAT, M. (org.). *Guerreiros sem espada: experiências revistas dos irmãos Villas Bôas*. São Paulo: Empresa das Artes, 1995.

BLOSSFELD, H. *Jardinagem*. São Paulo: Melhoramentos.

BONDAR, G. *A cultura do cacau na Bahia - Boletim Técnico nº 1 do Instituto de Cacau da Bahia*. São Paulo: Empreza Graphica da "Revista dos Tribunaes", 1938.

———. *Palmeiras do Brasil*. São Paulo: Instituto de Botânica/Secretaria de Agricultura do Estado de São Paulo, 1964

BOPP, R. *Cobra Norato e outros poemas*. Rio de Janeiro: Civilização Brasileira, 1978.

"BRASIL ganha briga na produção de cacau." *Diário de Pernambuco*, Diário Rural, Recife, 19 de outubro de 1990.

CALZAVARA, B.B.G. *Fruticultura tropical: a gravioleira (Annona muricata* L.*)*. Belém: EMBRAPA/Centro de Pesquisa Agroflorestal do Trópico Úmido, 1987.

———. *Recomendações básicas 1: cupuaçuzeiro*. Belém: EMBRAPA/Centro de Pesquisa Agroflorestal do Trópico Úmido, 1987.

———. *Recomendações básicas 3: açaizeiro*. Belém: EMBRAPA/Centro de Pesquisa Agroflorestal do Trópico Úmido, 1987.

———. *Recomendações básicas 4: cajueiro*. Belém: EMBRAPA/Centro de Pesquisa Agroflorestal do Trópico Úmido, 1987.

CARVALHO, A. *Histórico do desenvolvimento do cultivo do café no Brasil*. São Paulo: Instituto Agronômico de Campinas, 1993. Documentos IAC nº 34.

CASCUDO, L. da C. *História da alimentação no Brasil*. Belo Horizonte/São Paulo: Itatiaia/EDUSP, 1983. 2 volumes.

———. *Antologia da alimentação no Brasil*. Rio de Janeiro: Livros Técnicos e Científicos, 1977.

CAVALCANTE, P. B. *Frutas comestíveis da Amazônia*. Belém: Museu Paraense Emílio Goeldi/CNPq/INPA, 1972. volume 1.

———. *Frutas comestíveis da Amazônia*. Belém: Museu Paraense Emílio Goeldi/CNPq/INPA, 1974. volume 2.

———. *Frutas comestíveis da Amazônia*. Belém: Museu Paraense Emílio Goeldi/CNPq/INPA, 1979. volume 3.

CLEMENT, C. R. "Frutas da Amazônia". *Ciência Hoje*. São Paulo, agosto de 1992. p. 28-37

COELHO, Y. da S. *Lima ácida "tahiti" para exportação: aspectos técnicos da produção*. Brasília: EMBRAPA-SPI, 1993. Série Publicações Técnicas FRUPEX, nº 1.

COELHO, S. e RECCO, R. "A laranja chega a Goiás e Paraná". *Globo Rural*. São Paulo, março de 1989. p. 56-62.

COSTA, P. J. A. *A comida baiana de Jorge Amado ou O livro de cozinha de Pedro Arcanjo, com as merendas de Dona Flor*. São Paulo: Maltese, 1994.

COUCEIRO, E. M. "Cereja das antilhas, 7 colheitas por ano". *Ciência Hoje*. São Paulo, março/abril de 1985. p. 9.

COUTINHO, L. M. "O Cerrado e a ecologia do fogo." *Ciência Hoje*. São Paulo, Edição especial: ECO-Brasil. p. 130-138.

D'ÁVILA, C. *Boas maneiras: Manual de Civilidade*. Rio de Janeiro: Civilização Brasileira, 1936.

DIERBERGER, J. E. e MARINO NETTO, L. *Noz macadâmia. Uma nova opção para a fruticultura brasileira*. São Paulo: Nobel, 1985.

EMPRESA Brasileira de Pesquisa Agropecuária. *Programa Nacional de Pesquisa de Caju*. Fortaleza: EMBRAPA/Centro Nacional de Pesquisa de Caju, 1991.

FALCÃO, M. de A. *Aspectos fenológicos e de produtividade de algumas fruteiras cultivadas na Amazônia*. FUA/CNPq/INPA. Manaus, 1979.

FREIRE, G. *Açúcar. Em torno da etnografia, da história e da sociologia do doce no Nordeste canavieiro do Brasil*. Ministério da Indústria e do Comércio/Instituto do Açúcar e do Álcool, 1969. Coleção Canavieiras, número 2.

FREITAS, P. D. e SILVA, S. *Plantas brasileiras para a fitocosmética do século XXI. Um encontro de saúde, ciência, beleza e arte*. São Paulo: FCF-USP, 1995.

GAYET, J. P. *et alli*. *Lima ácida "tahiti" para exportação: procedimentos de colheita e pós-colheita*. Brasília: EMBRAPA-SPI, 1995. Série Publicações Técnicas FRUPEX, nº 12.

———. *Mamão para exportação: procedimentos de colheita e pós-colheita*. Brasília: EMBRAPA-SPI, 1995. Série Publicações Técnicas FRUPEX, nº 14.

GIRALDI, A. "Exemplo na floresta.". *Terra*. São Paulo, fevereiro de 1995, nº 34. p. 66-69.

GONÇALVES, J. A. "EMBRAPA pesquisa frutas nativas dos cerrados". *Folha de São Paulo*, Agrofolha, 20 de julho de 1993.

———. "Pesquisas melhoram a qualidade do maracujá." *Folha de São Paulo*, Agrofolha, 31 de janeiro de 1995.

GONSALVES, A. D. *O babaçu - considerações científicas, técnicas e econômicas*. Rio de Janeiro: Ministério da Agricultura/Serviço de Informação Agrícola, 1955.

GORGATTI NETTO, A. *et alli*. *Melão para exportação: procedimentos de colheita e pós-colheita*. Brasília: EMBRAPA-SPI, 1994. Série Publicações Técnicas FRUPEX, nº 6.

———. *Uva para exportação: procedimentos de colheita e pós-colheita*. Brasília: EMBRAPA-SPI, 1993. Série Publicações Técnicas FRUPEX, nº 2.

GUIMARÃES, O. "Sabores preservados". *Globo Rural*. São Paulo, março de 1992. p. 48-55.

GUIMARÃES ROSA, J. *Grande sertão: veredas*. Rio de Janeiro: José Olympio, 1970.

HOEHNE, F.C. *Frutas indígenas*. São Paulo: Instituto de Botânica de São Paulo,. 1946. 1ª reimpressão fac-similar: 1979.

INSTITUTO Brasileiro de Frutas. *Soluções fruta a fruta: acerola*. Edição nº 2. São Paulo: IBRAF, 1995.

———. *Soluções fruta a fruta: limão taiti*. Edição nº 3. São Paulo: IBRAF, 1995.

———. *Soluções fruta a fruta: manga*. Edição nº 1. São Paulo: IBRAF, 1995.

JORGE, J.A., LOURENÇÃO, A.L. e ARANHA, C. (editores). *Instruções agrícolas para o Estado de São Paulo*. Campinas: Instituto Agronômico de Campinas, 1994. 5ª ed.rev.atual.

JOLY, A. B. *Conheça a vegetação brasileira*. São Paulo: Polígono/EDUSP, 1970.

LEDO, A. da S. *Recomendações básicas para o cultivo da gravioleira (Annona muricata* L.*)*. Rio Branco: EMBRAPA/Centro de Pesquisa Agroflorestal do Acre, 1992.

LEITE, M. "Nova Califórnia renasce com o cupuaçu." *Folha de São Paulo*, 19 de dezembro de1993.

LORENZ, S. da S. *Sateré-Mawé: os filhos do guaraná*. São Paulo: Centro de Trabalho Indigenista, 1992.

LORENZI, H. *Árvores brasileiras - manual de identificação e cultivo de plantas arbóreas nativas do Brasil*. Nova Odessa: Plantarum, 1992.

LOUREIRO, C. *Doces lembranças*. São Paulo: Marco Zero, 1988.

LOURENÇO, M. "Baru: ficha da planta". *Globo Rural*. São Paulo, setembro de 1995. p.72-75.

———. "O sabor amargo do campo". *Globo Rural*. São Paulo, fevereiro de 1996. p.37-41.

MARCELINO, J. B. O. *Noções Preliminares sobre o cultivo do mangostão - Garcinia mangostana* L. Ilhéus: Comissão Executiva do Plano da Lavoura Cacaueira/CEDEX, 1990.

MARCONI, M. de A. *Folclore do café*. São Paulo: Secretaria da Cultura, Ciência e Tecnologia, Conselho Estadual de Cultura, 1976.

MARTINS, F. P. e PEREIRA, F. M. *Cultura do Caquizeiro*. Jaboticabal: FUNEP/UNESP, 1989.

MATTOS, J. R. "Frutos indígenas e seu estudo químico". *Almanaque Agrícola Chácaras e Quintais*, julho de 1961. p.131-159.

MEDEIROS, A. R. M. *A cultura da figueira*. Pelotas: EMBRAPA/Centro Nacional de Pesquisa de Fruteiras de Clima Temperado, 1987. Circular Técnica nº 13.

MELETTI, L.M.M. e SOUZA, J.S.I. de. *Maracujá: espécies, variedades e cultivo*. Piracicaba: Fundação de Estudos Agrônomos "Luiz de Queiróz", 1996.

MERCEDES-BENZ do Brasil. *Pindorama*. São Bernardo do Campo: Cartaz Editorial, 1993.

MONTEIRO, S. e KAZ, L. (orgs.). *Cerrado - vastos espaços*. Rio de Janeiro: Alumbramento/Livroarte, 1992-93.

——— (orgs.) *Sertão - Sertanejos*: Rio de Janeiro: Alumbramento/Livroarte, 1994-95.

MONTEIRO LOBATO, J. B. *Obras completas de Monteiro Lobato: Caçadas de Pedrinho e Hans Staden*. São Paulo: Brasiliense, 1966. Volume 3.

———. *Obras completas de Monteiro Lobato: Os doze trabalhos de Hércules (2º tomo)*. São Paulo: Brasiliense, 1966. Volume 17.

MOREIRA, J. C. "A ruína dos coronéis." *Veja*. São Paulo, 13 de setembro de 1995.

MOTA, M. *O cajueiro nordestino*. Recife: Fundação de Cultura Cidade de Recife, 1982.

MÜLLER, C.H. e CALZAVARA, B.B.G.. *Recomendações básicas 11: castanha-do-brasil*. Belém: EMBRAPA/Centro de Pesquisa Agroflorestal do Trópico Úmido, 1989.

——— *et alli*. *A cultura da castanha-do-brasil*. Brasília: EMBRAPA/Centro de Pesquisa Agroflorestal da Amazônia Oriental, 1995.

MUSEU Paraense Emílio Goeldi. *Cartilha informativa sobre a palmeira açaí (Euterpe oleracea* Mart.*)*. Belém: MPEG, 1995.

NOGUEIRA, O. L. *et alli*. *A cultura do açaí*. Brasília: EMBRAPA/Centro de Pesquisa Agroflorestal da Amazônia, 1995.

OLAYA, C. I. *Frutas de América - tropical y subtropical - historia y usos*. Santa Fé de Bogotá: Editorial Norma, 1991.

PIJPERS, D., CONSTANT, J.G. e JANSEN, K. *The complete book of fruit*. Nova Iorque: Gallery Books, 1986.

Publisher
Fábio Ávila

Editorial Direction
Heloisa Cavalcanti de Albuquerque

Supervision
Silvestre Silva

Graphic Design
Yvonne Sarué

Photography
Silvestre Silva

Text
Helena Tassara

Scientific Data
Sérgio Romaniuc Neto

Translation
Stuart C. Birkinshaw

Proofreading
Ronald Turnbull

Desktop Publishing
Fernando Barreto

Graphic Production
Marcos Albertin – Oficina de Arte
Renan Figueiredo

Color Separations
Gráfica Stampato

Printing and Binding
Laborgraf Artes Gráficas

Printed in Brazil
1998

Dados Internacionais de Catalogação na Publicação (CIP)
(Câmara Brasileira do Livro, SP, Brasil)

Silva, Silvestre P.
Fruit in Brazil / by Silvestre Silva. --
São Paulo : Empresa das Artes. 1996

1. Frutas - Brasil I. Título.

ISBN 85 - 85628 - 20 - 0

95-4368 CDD-634.0981

Índice para catálogo sistemático:
1. Brasil : Frutas 634.0981

EMPRESA DAS ARTES
Rua da Consolação, 2.851 - Cerqueira César
São Paulo - SP - Cep 01416-001
Fone (011) 853-3599 - 280-9658 - 881-3949